AF348494

A Systematic Approach to Electrodynamics

Rameez Ahmad Parra · Farooq Ahmad Dar ·
Mir Waqas Alam · Imtiyaz Ahmad Najar

A Systematic Approach to Electrodynamics

 Springer

Rameez Ahmad Parra
Department of Physics
University of Kashmir
Srinagar, Jammu and Kashmir, India

Mir Waqas Alam
Department of Physics
College of Science
King Faisal University
Ahsaa, Saudi Arabia

Farooq Ahmad Dar
Department of Physics
Central University of Kashmir
Tulmulla, Jammu and Kashmir, India

Imtiyaz Ahmad Najar
Department of Physics
Government Degree College
Charar-i-Sharief
Charar-i-Sharief, Jammu and Kashmir
India

ISBN 978-981-96-3407-1   ISBN 978-981-96-3408-8   (eBook)
https://doi.org/10.1007/978-981-96-3408-8

Deanship of Scientific Research, Vice Presidency for Graduate Studies and Scientific Research, King Faisal University, Saudi Arabia Project No. KFU251637.

Preface

Notwithstanding with the monumental success and strides made by the Newtonian mechanics in the first quarter of the twentieth century, it was found to be inadequate to explain the plethora of problems wherein either the objects move with exceptionally very high velocity or the size of the object is extremely small. The problems involving objects moving with velocities comparable to the velocity of light are best explained in the realm of theory of special relativity propounded by Einstein in 1905. The fundamental principle of special theory of relativity can be enunciated as, "Physical law is independent of the inertial frame of reference; i.e., a physical law maintains its structure in all inertial reference frames." This fundamental assumption is often dubbed as covariance of physical law. For instance, we consider the velocity of light in vacuum to be a physical law; i.e., the velocity of light remains constant in all inertial reference frames. Further, the negative results of Michelson and Morley experiment were explained by Lorentz and Fitzgerald. These mathematicians changed the notion of space and time considered under Newtonian mechanics and thereby compelled Einstein to formulate a new concept of spacetime entity. This spacetime is a geometric framework within which we perform physics. Einstein ruled out the concept of ether and pronounced that the physical laws are unaffected by the motion of the observer. Moreover, the clear distinction between the classical theory and Einstein's theory of relativity lies in the conceptualization that the velocity of light in vacuum is unaltered by the velocity of observer or by the velocity of source. Further, a classical treatment is ruthlessly inadequate to explain the problems involving objects having extremely small sizes. These problems may fall within the domain of quantum mechanics, put forward by Schrodinger, Bohr, Heisenberg et al. in the early years of the twentieth century. Quantum mechanics describes a broad range of problems based on very few postulates. It provides a quantitative prediction for many physical situations, and these predictions agree well with the experiments thereof. To cut the long story short, quantum mechanics is the ultimate basis today by which we understand the physical world around us. A theory that subsumes relativity and quantum mechanics is referred as quantum field theory. However, even as of today it cannot contend to be a complete satisfactory theory. In the present context, we would exclusively work within the domain of electrodynamics. It is pertinent to note here that the electrodynamics is a branch of physics in which the force acting between two moving charges depends upon the distance between them and their velocities. Furthermore, the force is not directed along a straight line between the charges. This is the fundamental difference between classical mechanics and electrodynamics. Therefore, the basic concepts of classical mechanics are not applicable to electrodynamics. The extension of the principle of relativity to electrodynamics results in the development of four fundamental Maxwell's equations for empty space. According to Maxwell, "Electromagnetic waves propagate in vacuum with a uniform velocity $c = 3 \times 10^8$ m/s. Light waves are electromagnetic waves, and the velocity of light in vacuum is unaltered by the state of motion of the source of light."

The present book has been devised for graduate students and can be covered comfortably in one semester. We have tried our best to make complex and intriguing concepts and ideas more interesting and accessible. The book is divided into five chapters. Every chapter is supplemented with problems besides solved examples related to the topics covered. Some of the problems are more challenging, and therefore, a reader has to render a lot of effort to solve such problems. However, it is worthwhile that such problems will help the reader to fathom the beauty of the subject. Chapter 1 deals with the basic mathematical ideas and concepts. A strong mathematical background is prerequisite to better understand this subject, and therefore, the readers are advised to devote a fair amount of time on this chapter. Moreover, the readers should also benefit from other standard textbooks on mathematical physics so as to get the proper insight into the realm of mathematics. Chapter 2 of the book concerns with the boundary value problems. The main focus has been laid on solving the Laplace equation under suitable boundary constraints in spherical and cylindrical coordinates. The end results are most profound and thought provoking. The readers will be benefitted from the present book as most of the derivations have been carried out in a diligent manner. The readers are advised to emphasize on the problems of this chapter. Chapter 3 illustrates the intricate boundary value problems. The boundary value problems based on the method of images have been covered in this chapter. This chapter will help the reader to get deep insight into the subject. Chapter 4 deals with the advanced topics of electrodynamics. This chapter illustrates the potential formulation of electric and magnetic fields. The advanced topics presented here are predominantly those involving the interaction of the charged particles with each other and with the electromagnetic fields. Chapter 5 is devoted to the study of relativistic particle kinematics and dynamics. Once the four-vector concept is properly understood, the reader will be in a better position to employ the Lorentz transformations in explaining various phenomena appearing in relativistic electrodynamics. The end results of this chapter are very interesting and cannot be found elsewhere in the literature.

Srinagar, India   Rameez Ahmad Parra
Tulmulla, India   Farooq Ahmad Dar
Ahsaa, Saudi Arabia   Mir Waqas Alam
Charar-i-Sharief, India   Imtiyaz Ahmad Najar

Acknowledgements This work was supported by the Deanship of Scientific Research, Vice Presidency for Graduate Studies and Scientific Research, King Faisal University, Saudi Arabia, under Project No. KFU251637. We would like to express our sincere gratitude for their support and assistance throughout this work. We would also like to express our sincere gratitude to a number of friends and colleagues whose suggestions, questions and comments have benefitted us a lot and without their persistent interest in our work, this book would never have been completed. It would be difficult to list all those friends and colleagues one by one. We are indebted to their insightful perception, illuminating advices and continuous encouragement during the entire work. Moreover, we would like to extend our deep regards and heartfelt appreciation to our friend, viz. Er. Yehya Rasool, dual Ph.D. scholar at IIT Kanpur and NYCU Taiwan, who rendered us immense help and contributed a lot for the completion of this strenuous task. The open-hearted cooperation of Er. Yehya Rasool is unfathomable. He is credited to have designed all the diagrams in this book.

Funding Statement The authors declare the following as potential competing interests: This work was supported by the Deanship of Scientific Research, Vice Presidency for Graduate Studies and Scientific Research, King Faisal University, Saudi Arabia, under Project No. KFU251637 wmir@kfu.edu.sa.

Fundamental Interactions

According to our current understanding, nature is governed by four fundamental interactions: gravitational, weak, electromagnetic and strong. The relative strength and range of these interactions vary significantly. However, physicists have embarked on a strenuous task to search for the possibility of unification of these fundamental interactions and that would serve as a strong tool to develop the concise description of the Universe. After persistent development, electromagnetic and weak interactions were dubbed as the two facets of the electroweak force. This theory was proposed by Glashow, Weinberg and Salam. This discovery was important in formulating the Standard Model of particle physics. This model was successfully employed to explain a wide variety of problems with the utmost precision. This model provides a coherent quantum-mechanical description of electromagnetic, strong and weak interactions based on the fundamental constituents of matter, viz. quarks and leptons interacting via force carriers' photons, $W^{\pm}$ and Z bosons and gluons. However, electroweak theory is still considered incomplete as it sans in providing any viable explanation in the asymmetry between matter and antimatter. The theory that describes the characteristic features of the strong force is referred as Quantum Chromodynamics (QCD). The existence of the disentangled state of matter called Quark Gluon Plasma (QGP) was predicted by the QCD. QCD is a gauge theory; i.e., it is based on the fundamental postulate that if all colours in a system are simultaneously changed, this does not affect the interactions within the system. A principle often referred as gauge invariance or the invariance under gauge transformations. However, systems in QCD are even invariant under gauge transformations that depend on positions in space and time. This assumption is referred as local gauge invariance, and thus QCD is called a local gauge theory. The electromagnetic force is described by the Abelian field theory called Quantum Electrodynamics (QED) and is based on symmetry group $U(1)$. This implies the existence of the electric charge and one colourless field, the photon. However, the symmetry group that describes the strong interactions is the non-Abelian group $SU(3)$ that implies the existence of eight massless gluon fields (associated with eight Gell-Mann matrices, generators of the $SU(3)$ group) which carry the colour charge and are, therefore, auto-interacting.

Contents

About the Editors

Dr. Rameez Ahmad Parra began his academic journey at Aligarh Muslim University, where he completed his B.Sc. (Hons). and M.Sc. in Physics. He then pursued a Ph.D. in High Energy Physics at Jamia Millia Islamia, New Delhi, making valuable contributions to the field. He has held teaching positions at Central University of Kashmir and Cluster University Srinagar and currently works as an Assistant Professor (C) at the University of Kashmir, Srinagar. He has taught a diverse range of subjects, including Statistical Mechanics, Classical Electrodynamics, Particle Physics, Neutrino Physics, and Radiation Physics. In addition to teaching, Dr. Parra is actively involved in research, with his work published in well-regarded journals and presented at national and international conferences, reflecting his commitment to fostering education and driving scientific progress.

Dr. Farooq Ahmad Dar began his studies in Physics at the University of Kashmir and went on to earn his Master's degree with distinction (topper) from Dr. Harisingh Gour Central University, Sagar, Madhya Pradesh. His academic pursuit culminated in a Ph.D. from the National Institute of Technology (NIT) Srinagar, where his pioneering research on nanomaterials significantly advanced applications in energy storage, sensors, and dielectric materials. Currently serving as an Assistant Professor (c) in the Department of Physics at the Central University of Kashmir, Dr. Dar is actively involved in teaching, mentoring, and research. His areas of expertise include nanomaterials, thin films, and doped metal oxides with applications in electronics, optics, and energy conservation, and his work has been widely published in reputed scientific journals. In addition to his academic contributions, Dr. Dar is committed to community outreach, particularly in promoting science education among underrepresented groups. He believes deeply in the transformative power of education and strives to foster curiosity, critical thinking, and a love for learning in both students and the wider community, embodying a vision of science that is accessible, impactful, and inclusive.

Dr. Mir Waqas Alam is a distinguished scientist and Associate Professor renowned for his expertise in Nano and Functional Material Science. He currently serves in the Department of Physics at the College of Science, King Faisal University, Saudi Arabia, where he was appointed as an Assistant Professor in September 2015. Recognizing his exceptional contributions and dedication, he was subsequently promoted to the rank of Associate Professor in October 2022. Dr. Alam's academic journey has been marked by a relentless pursuit of knowledge and innovation. He earned his Ph.D. in Nano and Functional Material Science from Toyama University, Japan, where he delved deep into the intricacies of his field. His quest for excellence led him to various esteemed institutions worldwide, including Western Kentucky University, USA, where he served as a research scholar, and the National Institute of Material Science (NIMS), Japan, where he worked as a post-doctoral fellow.

Throughout his career, Dr. Alam has demonstrated a remarkable breadth of research interests and accomplishments. His research is centered around Nano and Functional Material Science, with a particular emphasis on the fabrication and characterization of organic and photonic devices. His expertise encompasses a wide array of cutting-edge technologies, including Organic Thin-Film Transistors (OTFTs), Organic Optically Functional OFETs, and various energy devices such as solar cells. His work delves into the multifaceted applications of nanotechnology, ranging from antimicrobial and antioxidant properties to bio-related applications. Through meticulous experimentation and thorough characterization techniques, Dr. Alam seeks to uncover novel functionalities and applications for nanomaterials, pushing the boundaries of scientific innovation.

Dr. Alam's commitment to excellence is reflected not only in his research endeavors but also in his dedication to education and mentorship. Over the past five years, he has published over 70 high-impact factor journal papers and successfully completed several funded projects. Additionally, he has played a pivotal role in shaping the next generation of scientists by supervising numerous master's students in their thesis work.

Dr. Mir Waqas Alam's unwavering dedication to advancing scientific knowledge, coupled with his passion for mentorship and education, makes him a highly respected figure in the scientific community. His contributions have not only enriched the field of Nano and Functional Material Science but also inspired countless individuals to pursue excellence in their own scientific endeavors.

Dr. Imtiyaz Ahmad Najar completed his post-graduation in Physics from the Department of Physics, University of Kashmir, Srinagar. He pursued his Ph.D. in High Energy Physics from the Department of Physics, University of Kashmir, Srinagar.

During his doctoral studies, he has presented his research work at several national-level workshops and conferences. Further, he attended "The Nineth International School for Strangeness Nuclear Physics at J-PARC, Tokai, Japan" where he presented some part of his research work. He has published numerous papers in internationally reputed journals, with further work currently under publication. Apart from his research, he has several years of teaching experience, having served in various Colleges across Jammu & Kashmir. He has taught a wide range of subjects during his teaching tenure. Besides teaching, he is actively involved in research work which involves deciphering the properties of an extended phase of matter called as QGP.

Chapter 1
Mathematical Tools for Electrodynamics

Abstract This chapter introduces vectors as mathematical objects characterized by both magnitude and direction, distinguishing them from scalars. Concepts include vector notation, components and operations such as addition, scalar multiplication, dot and cross products and triple products, which are pivotal in physics and engineering. Graphical interpretation using arrows allows an understanding of these operations. Advanced topics cover transformations of vectors under coordinate changes, using tools like transformation matrices, Kronecker delta and Levi–Civita symbols. The discussion extends to the differential calculus of vectors, exploring gradient, divergence, curl and integral formulations (line, surface and volume integrals) critical for analysing physical fields. Special functions like Legendre polynomials and Bessel functions are introduced for problems involving spherical and cylindrical symmetries. The chapter concludes with the applications of vector concepts and specialized functions in electrodynamics, offering a foundation for analysing spatial and directional phenomena.

Keywords Scalar and vector · Tensor · Del operator · Levi–Civita symbol · Kronecker delta · Special functions

1.1 Significance of Mathematical Tools in Electrodynamics

Electrodynamics, the branch of physics that deals with the study of electric and magnetic fields, is inherently mathematical. This discipline explores how electric charges produce and interact with electric and magnetic fields, leading to various phenomena such as electromagnetism, electromagnetic induction and radiation. A strong mathematical formulation is, therefore, essential for understanding and predicting these phenomena.

© The Author(s), under exclusive license to Springer Nature Singapore Pte Ltd. 2025
R. Ahmad Parra et al., *A Systematic Approach to Electrodynamics*,
https://doi.org/10.1007/978-981-96-3408-8_1

1.2 Mathematical Formulation of Electrodynamics

1.2.1 Maxwell's Equations

These are unparalleled equations of electrodynamics. These constitute a set of differential equations that describe how electric and magnetic fields are generated and are altered by the charges and currents. These equations, expressed in differential form, are given as:

1. Gauss's Law: $\vec{\nabla} \cdot \vec{E} = \frac{\rho}{\varepsilon_0}$
2. Gauss's Law for Magnetism: $\vec{\nabla} \cdot \vec{B} = 0$
3. Faraday's Law: $\vec{\nabla} \times \vec{E} = -\frac{\partial \vec{B}}{\partial t}$
4. Ampere-Maxwell Law: $\vec{\nabla} \times \vec{B} = \mu_0 \vec{J} + \mu_0 \varepsilon_0 \frac{\partial \vec{E}}{\partial t}$

These equations are inherently mathematical and provide a comprehensive framework for understanding electromagnetic phenomena. To unlock their full predictive power, a range of mathematical tools is essential. These tools help physicists analyse electromagnetic systems and solve boundary value problems.

1.2.2 Vector Calculus

Electromagnetic fields are vector fields. Understanding concepts like divergence, gradient and curl are fundamental to vector calculus and are, therefore, essential for interpreting and applying Maxwell's equations.

1.3 Quantitative Predictions and Analysis

Mathematical tools enable the precise quantification of electromagnetic phenomena. Calculations involving electric field strengths, magnetic flux and electromagnetic wave propagation require numerical methods and analytical skills. Complex phenomena like wave interference, polarization and diffraction can be accurately modelled and predicted using mathematical equations.

1.4 Computational Electrodynamics

Advancements in computational methods make mathematical tools crucial for simulating electromagnetic fields and waves in complex systems. Techniques like Finite Element Analysis (FEA) are extensively used in designing and testing electronic devices, antennas and other systems. Computational models rely on numerical solutions to Maxwell's equations, enabling scientists and engineers to visualize and optimize electromagnetic interactions in real-world scenarios.

1.5 Applications

Undoubtedly mathematical tools form the bedrock of modern-day science and technology. The various applications of such tools can be illustrated as follows.

1.5.1 Engineering and Technology

Mathematical tools in electrodynamics are important for designing various technological applications, from day-to-day gadgets to sophisticated communication systems and power grids. Understanding electromagnetic compatibility and interference, essential for the functioning of electronic devices, relies heavily on mathematical analysis.

1.5.2 Research and Innovation

Advanced mathematical techniques in electrodynamics pave the way for research in areas like photonics, quantum computing and wireless energy transfer. Theoretical predictions often lead to new discoveries and technological advancements. Thus, in short, we can say that mathematical tools are not just beneficial but essential in the field of electrodynamics. They provide the language and framework for understanding, predicting and applying the principles of electromagnetism. From theoretical research to practical engineering, the role of mathematics in electrodynamics is fundamental and far-reaching, underpinning the technological advancements that drive our modern world.

In the realm of physics, physical quantities are categorized into two distinct types: scalar and vector. Scalar quantities are simpler in nature and can be effectively managed using ordinary algebra due to their singular attribute of magnitude. Conversely, vector quantities are more complex, possessing both magnitude and direction, necessitating the use of a specialized form of algebra tailored for them. Our objective is to develop and explore this specialized algebra, specially designed for handling vector quantities.

1.6 Definition and Representation of a Vector

A vector is a mathematical object possessing both magnitude and direction. It follows vector laws that distinguish it from a scalar, which has only magnitude. Vectors are used in physics to represent quantities that need both these properties to be fully described.

1.6.1 Notation and Components

- A vector is often represented by an arrow over a symbol, for instance, $\vec{A}$. It is also written as a bold-faced letter **A.**
- The size or magnitude of the vector $\vec{A}$ is represented by $\left|\vec{A}\right|$ or, simply by A and refers to the length or size of the vector.
- Generally, a vector of unit magnitude drawn in the direction of a given vector represents the direction of a given vector. Therefore, for any vector $\vec{A}$, we represent a unit vector by $\hat{A}$.

1.6.2 Vector Representation

- The vector $\vec{A}$ can be expressed as a product of its magnitude and its direction: $\vec{A} = \left|\vec{A}\right|\hat{A}$.
- In a Cartesian coordinate system, vectors can be expressed into components along the three coordinate axes. For instance,

$$\vec{A} = A_x\hat{i} + A_y\hat{j} + A_z\hat{k} \tag{1.1}$$

where $\hat{i}, \hat{j}$ and $\hat{k}$ are orthogonal unit vectors.

1.6.3 Negative of a Vector

- The negative of a vector $\vec{A}$ is symbolized as $-\vec{A}$. It has essentially the same magnitude as $\vec{A}$; however, it is directed opposite to vector $\vec{A}$.
- This can be represented as $-\vec{A} = \left|\vec{A}\right|(-\hat{A})$. Here $-\hat{A}$ signifies a unit vector in the direction opposite to $\hat{A}$.

Fig. 1.1 Negative vector

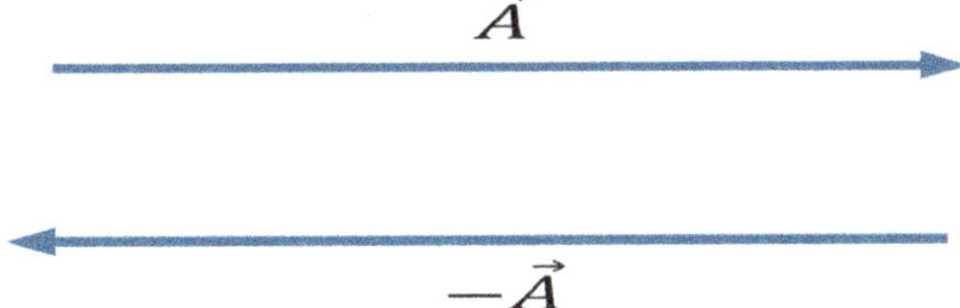

1.6.4 Graphical Interpretation

- Graphically, if you draw a vector $\vec{A}$ as an arrow pointing in a certain direction, then $-\vec{A}$ will be an arrow of the same length pointing in the exactly opposite direction, as shown in Fig. 1.1.
- This concept is fundamental in vector algebra for operations like vector subtraction, where the negative of a vector is often utilized.

1.6.5 Addition of Vectors

Vector addition is a fundamental operation in vector algebra and is distinct from the addition of scalar quantities. The process involves combining two or more vectors to form a single vector, known as the resultant vector. Let us elaborate on the vector addition process using the example of adding vectors $\vec{A}$ and $\vec{B}$ to form the resultant vector $\vec{C}$ represented as $\vec{A} + \vec{B} = \vec{C}$ as shown in Fig. 1.2.

Fig. 1.2 Vector addition

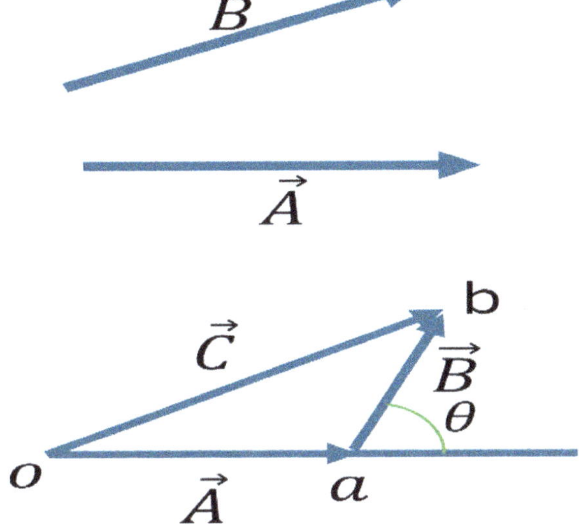

1.6.6 Graphical Representation

The method of vector addition is illustrated as follows:

1. **Head-to-Tail Method**:

   - To add $\vec{A}$ and $\vec{B}$, we first draw vector $\vec{A}$ with its tail at the origin and its head pointing in its designated direction.
   - Then, translate vector $\vec{B}$ so that its tail is at the head of $\vec{B}$. The translation of vector $\vec{B}$ should be such that its magnitude and direction remain unchanged.
   - The resultant vector $\vec{C}$ is then drawn from the tail of $\vec{A}$ (the starting point) to the head of $\vec{B}$ (the end point).

2. **Translation of Vectors**:

   - In vector addition, it's important to note that the vectors can be moved (translated) parallel to themselves in space without changing their effect. This property is crucial for vector addition.

3. **Resultant Vector**:

   - The resultant vector $\vec{C}$ in this case represents the combined effect of vectors $\vec{A}$ and $\vec{B}$. Its magnitude and direction are determined by the combined effect of the two original vectors.

1.6.7 Commutative Nature

Vector addition always follows the commutative law.

   In Fig. 1.3, vector $\vec{A}$ is drawn along OC with its head pointing in a certain direction. Vector $\vec{B}$ then placed with its tail at the head of $\vec{A}$ pointing in its own direction. The resultant vector $\vec{C}$ is drawn from the tail of $\vec{A}$ to the head of $\vec{B}$.

   Here is the illustration of vector addition using the head-to-tail method. In this graphical representation:

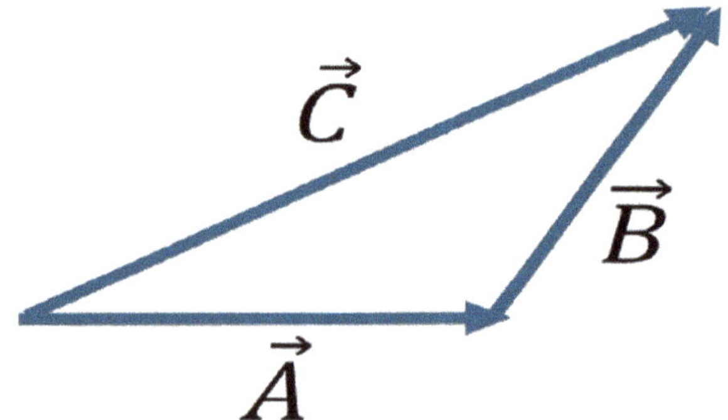

Fig. 1.3 Graphical representation of vector addition, illustrating the resultant vector obtained by combining two vectors using the head-to-tail method

- Vector $\vec{A}$ is drawn from the origin with its head pointing in a specific direction.
- Vector $\vec{B}$ is then placed with its tail at the head of $\vec{A}$ pointing in its own direction.
- The resultant vector $\vec{C}$ is depicted from the tail of $\vec{A}$ to the head of $\vec{B}$ representing the combined effect of vectors $\vec{A}$ and $\vec{B}$.

1.7 Multiplication of Vectors

Vector multiplication can be performed in different ways depending on whether the multiplier is a scalar or another vector. This process is fundamental in physics and engineering for various applications.

1.7.1 Multiplication of Vector by a Scalar

When a vector quantity is multiplied by a scalar quantity, then the new physical quantity obtained is also a vector quantity. The magnitude of this new physical quantity is equal to the product of the scalar quantity and the magnitude of the vector quantity. However, the direction of the new vector quantity is same as that of the vector quantity with which a scalar quantity is multiplied. For example, if $\vec{A}$ is a vector and 3 is a scalar, then $3\,\vec{A} = 3\left|\vec{A}\right|\hat{A}$. Here, the magnitude of vector $\vec{A}$ is multiplied by 3, but its direction remains unchanged.

1.8 Vector Multiplication

Vector multiplication is of two types, viz. scalar product and vector product.

1.8.1 Scalar Product

It is also called as dot product.

- The scalar or dot product of two vectors results in a scalar quantity.
- Representation: $\vec{A}.\vec{B}$.

Geometrical Interpretation: The dot product measures the extent to which two vectors align with each other. Mathematically, the dot product of two vectors $\vec{A}$ and $\vec{B}$ is defined as the product of their magnitudes multiplied by the cosine of the smaller angle between them

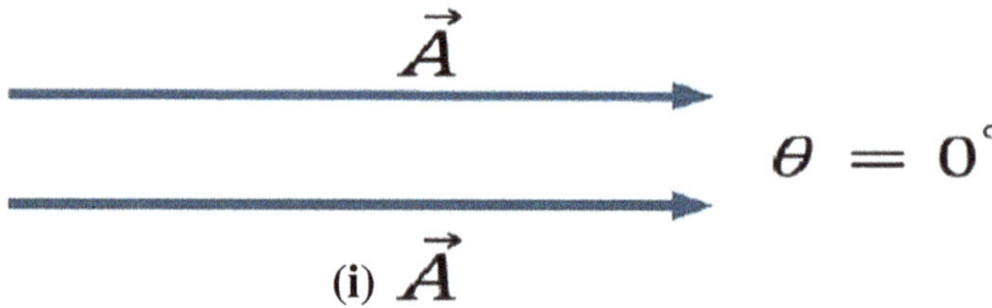

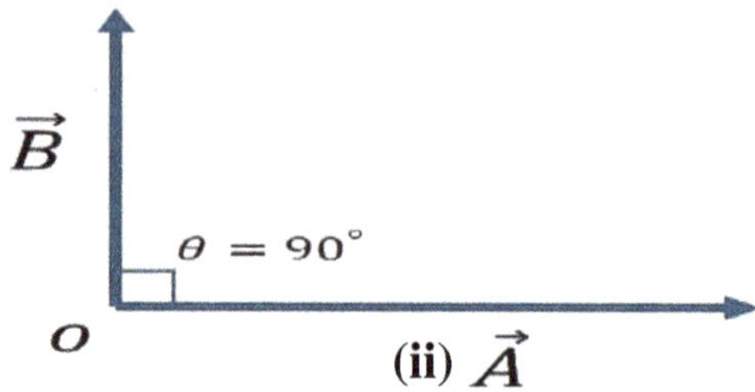

Fig. 1.4 Two vectors at $\theta = 0^0$ and 90^0. Area $= ab$. On the other hand, $\vec{A}.\vec{B} = 0$ (at $\theta = 90^0$)

$$\vec{A}.\vec{B} = \left|\vec{A}\right|\left|\vec{B}\right|\cos(\theta) \tag{1.2}$$

- **Commutativity**: It follows commutative law, i.e., $\vec{A}.\vec{B} = \vec{B}.\vec{A}$
- **Special Cases**: If the angle between the vectors is 0°, then $\vec{A}.\vec{A} = A^2$. However, if the angle between vectors is 90°, then $\vec{A}.\vec{B} = \vec{B}.\vec{A} = 0$ indicating that the vectors are perpendicular, hence no projection of one on the other as shown in Fig. 1.4.

In physics, the dot product is used to calculate work done: $W = \vec{F} \cdot \vec{ds}$, where $\vec{F}$ is the force vector and $\vec{ds}$ is the displacement vector.

1.8.2   Law of Cosines

The dot product concept leads to the law of cosines in geometry, which is useful in solving problems involving angles and lengths in a triangle (Fig. 1.5).

To develop a better understanding of dot product, consider the example.

Example 1.1 Given the vector equation $\vec{B} + \vec{C} = \vec{A}$. Find $\vec{C}.\vec{C}$.

Solution:

From the above we can write:

$$\vec{C} = \vec{A} - \vec{B} \tag{1.3}$$

Fig. 1.5 The geometry of
vector addition

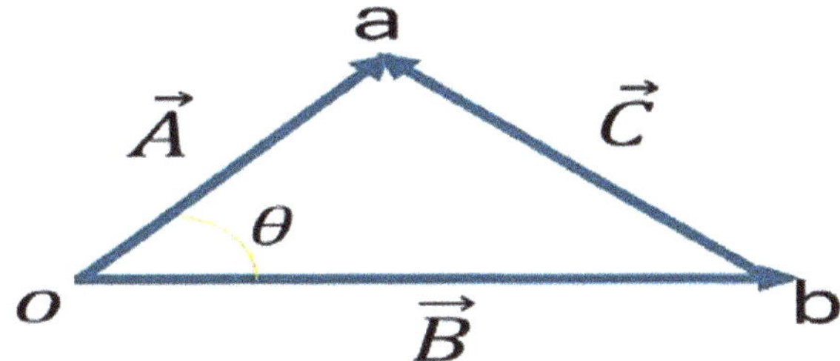

Taking the dot product of this expression with itself, we obtain (Fig. 1.5).

$$\vec{C} \cdot \vec{C} = \left(\vec{A} - \vec{B}\right) \cdot \left(\vec{A} - \vec{B}\right)$$

$$\vec{C} \cdot \vec{C} = \vec{A} \cdot \left(\vec{A} - \vec{B}\right) - \vec{B} \cdot \left(\vec{A} - \vec{B}\right)$$

$$\vec{C} \cdot \vec{C} = \vec{A} \cdot \vec{A} - \vec{A} \cdot \vec{B} - \vec{B} \cdot \vec{A} + \vec{B} \cdot \vec{B}$$

$$C^2 = A^2 - 2\vec{A} \cdot \vec{B} + B^2$$

$$C^2 = A^2 + B^2 - 2\vec{A} \cdot \vec{B}$$

$$C^2 = A^2 + B^2 - 2\left|\vec{A}\right|\left|\vec{B}\right|\cos(\theta) \tag{1.4}$$

Similarly, we can prove that

$$B^2 = A^2 + C^2 - 2\left|\vec{A}\right|\left|\vec{C}\right|\cos(\theta) \tag{1.5}$$

Proceeding in the same manner, we can show that

$$A^2 = B^2 + C^2 - 2\left|\vec{B}\right|\left|\vec{C}\right|\cos(\theta) \tag{1.6}$$

Thus, Eqs. (1.4), (1.5) and (1.6) are known as law of cosines. This example
demonstrates how the dot product is used in vector algebra to solve complex problems
involving vector magnitudes and angles.

Example of dot product is work: One of the most important applications of scalar
or dot product in physics is in the calculation of work, $W = \vec{F} \cdot \vec{ds}$, which is interpreted
as displacement times the projection of the force along the direction of displacement
vector.

Example 1.2 A force $\vec{F} = 5\hat{i} + 2\hat{j} - \hat{k}N$ acts on an object that undergoes a
displacement $\vec{d} = 4\hat{i} - \hat{j} + 2\hat{k}m$. What is the work done by the force?

Solution:

Work performed by the force is given by

$$W = \vec{F}.\vec{d}$$

$$W = 16\,\mathrm{J}$$

The concept of dot product naturally leads to the Law of Cosines in geometry, which is a fundamental tool in solving problems involving angles and length of sides in a triangle. The Law of Cosines is particularly useful in cases where the Pythagoras theorem does not apply, such as nonright-angled triangles. This law establishes a relationship between the sides of a triangle and the cosine of one of its angles.

Mathematically, the Law of Cosines states that

$$C^2 = A^2 + B^2 - 2\left|\vec{A}\right|\left|\vec{B}\right|\cos(\theta)$$

Where

- A and B are the lengths of two sides of the triangle,
- θ is the angle between these two sides,
- C is the length of the side opposite to the angle θ.

Geometric Interpretation

- The Law of Cosines is a generalization of the Pythagoras theorem. When $\theta = 90^0$, $\cos(90^0) = 0$ and the equation simplifies to the familiar form $C^2 = A^2 + B^2$ which is the Pythagoras theorem often encountered in mathematics. However, when $\theta \neq 90^0$, the additional term $-2\left|\vec{A}\right|\left|\vec{B}\right|\cos(\theta)$ accounts for the influence of the angle on the length of the third side.

Visual Representation

Figure 1.4: Two vectors at different angles

- (i) When $\theta = 0^0$, the two vectors are aligned in the same direction, maximizing their dot product.
- (ii) When $\theta = 90^0$, the dot product is zero because $\cos(90^0) = 0$, meaning the vectors are perpendicular.

By understanding the relationship between the dot product and the Law of Cosines, one can gain deeper insights into both vector algebra and geometric principles, making it easier to solve complex problems in various scientific and engineering fields.

1.8.3 Vector Product

It is also referred as cross product. The vector product of two vectors $\vec{A}$ and $\vec{B}$ is defined as the product of their magnitudes multiplied by the sine of the smaller angle

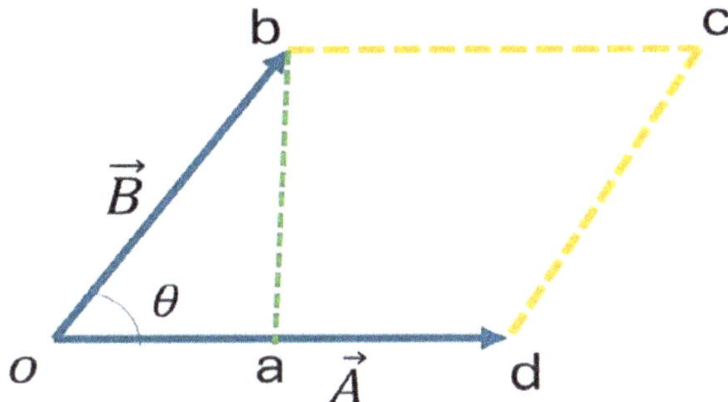

Fig. 1.6 Parallelogram representation of vector addition

between them. The direction of the resultant vector $\vec{C}$ is perpendicular to the plane containing vectors $\vec{A}$ and $\vec{B}$. Therefore, the vectors $\vec{A}$, $\vec{B}$ and $\vec{C}$ form a right-handed system. For instance, angular momentum and torque are examples of cross product (Fig. 1.6).

The cross product, also known as the vector or outer product, is a binary operation on two vectors in three-dimensional space. It results in a vector that is perpendicular to both of the original vectors and its magnitude is proportional to the area of the parallelogram that the vectors span.

1.8.4 Definition and Properties

1. **Mathematical Representation:**

   - The cross or vector product of two vectors $\vec{A}$ and $\vec{B}$ is denoted as $\vec{A} \times \vec{B}$.
   - It is given by $\vec{A} \times \vec{B} = |\vec{A}||\vec{B}| \sin(\theta)\hat{n}$ where θ is the angle between vectors $\vec{A}$ and $\vec{B}$ and $\hat{n}$ is a unit vector perpendicular to the plane containing $\vec{A}$ and $\vec{B}$.

2. **Non-commutative Nature**

   - The cross product is not commutative: $\vec{A} \times \vec{B} = -\vec{B} \times \vec{A}$.
   - This property implies that by reversing the order of the vectors in the cross product changes the direction of the resultant vector.

3. **Self-cross Product:**

   - The cross product of a vector with itself is zero: $\vec{A} \times \vec{A} = 0$ (No area swept). This is because the angle between $\vec{A}$ and itself is $0°$ making $\sin(0°) = 0$.

4. **Distributive Property:**

   The cross product follows the distributive law: $\vec{A} \times \left(\vec{B} + \vec{C}\right) = \vec{A} \times \vec{B} + \vec{A} \times \vec{C}$.

1.8.5 Application in Rotational Dynamics

1. **Torque:**

   - A primary example of the cross product in physics is torque, denoted as $\vec{\tau}$
   - Torque is defined as $\vec{\tau} = \vec{r} \times \vec{F}$ where $\vec{r}$ is the position vector and $\vec{F}$ is the force applied.
   - The direction of $\vec{\tau}$ is normal to the plane formed by $\vec{r}$ and $\vec{F}$ indicating $\vec{\tau}$ is along the axis of rotation.

2. **Work versus Torque:**

   - Although the dimensional formulae of work and torque are same, they describe different physical concepts.
   - Work is associated with translational motion (i.e., the straight-line movement of an object from one position to another). It is concerned with how much force has been applied to move an object and how far the object has moved. And is a scalar product of force and displacement. In the context of energy, work represents the transfer of energy to or from an object via the application of force along a displacement. For instance, lifting a weight off the ground does work against gravity, increasing the object's gravitational potential energy.
   - Torque, on the other hand, is related to rotational motion (torque is associated with rotational motion—the spinning or turning movement of an object around a centre or axis). It describes the twisting effect a force has on an object, determining how effectively a force causes an object to rotate. Just as force causes an object to accelerate linearly, torque causes an object to acquire angular acceleration. It is a pivotal concept in understanding rotational dynamics in systems ranging from simple mechanical levers to complex machinery and motors and is a vector product of the position vector and force.

1.8.6 Geometric Interpretation

The cross product can be visualized as the area of the parallelogram formed by the two vectors. The direction of the resultant vector (given by the right-hand rule) is normal to the plane containing the original vectors. Let us create a diagram to visually represent the vector product of two vectors highlighting their geometric relationship and the direction of the resultant of vectors $\vec{A}$ and $\vec{B}$ (Fig. 1.7).

Here is the diagram illustrating the vector product of two vectors $\vec{A}$ and $\vec{B}$. In this representation:

- Vectors $\vec{A}$ and $\vec{B}$ are shown forming an angle θ between them.
- The resultant vector $\vec{C} = \vec{A} \times \vec{B}$ is depicted as perpendicular to the plane containing $\vec{A}$ and $\vec{B}$.
- The parallelogram formed by vectors $\vec{A}$ and $\vec{B}$ is illustrated to help visualize the area aspect of the cross product.

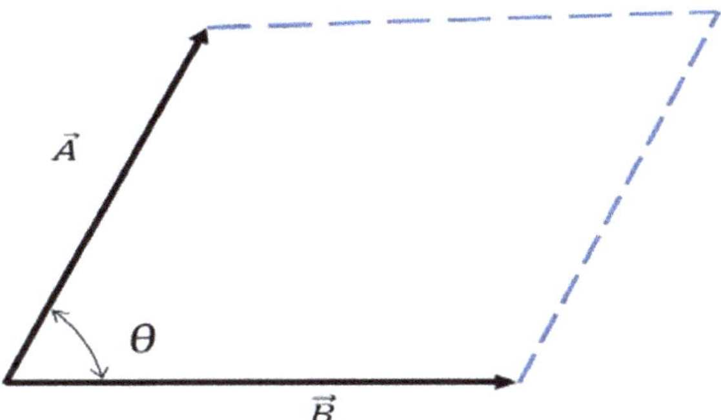

Fig. 1.7 Cross product in vector composition

- The direction of $\vec{C}$ is indicated in accordance with the right-hand rule.

Each vector is clearly labelled, and the angle θ is marked, providing a clear visual understanding of the geometric and directional properties of the cross product in vector algebra.

1.8.7 Vector Algebra

However, the three axes x, y and z are mutually perpendicular to each other and form a rectangular coordinate system. Let $\hat{x}$, $\hat{y}$ and $\hat{z}$ represent the unit vectors along x-, y- and z-axes, respectively. Therefore, we can write (Fig. 1.8):

$$\hat{x}.\hat{x} = \hat{y} \cdot \hat{y} = \hat{z}.\hat{z} = 1$$
$$\hat{x}.\hat{y} = \hat{y}.\hat{z} = \hat{z}.\hat{x} = 0$$
$$\hat{x} \times \hat{x} = \hat{y} \times \cdot\hat{y} = \hat{z} \times \hat{z} = 0$$
$$\hat{x} \times \hat{y} = \hat{z}, \hat{y} \times \hat{z} = \hat{x}, \hat{z} \times \hat{x} = \hat{y} \qquad (1.7)$$

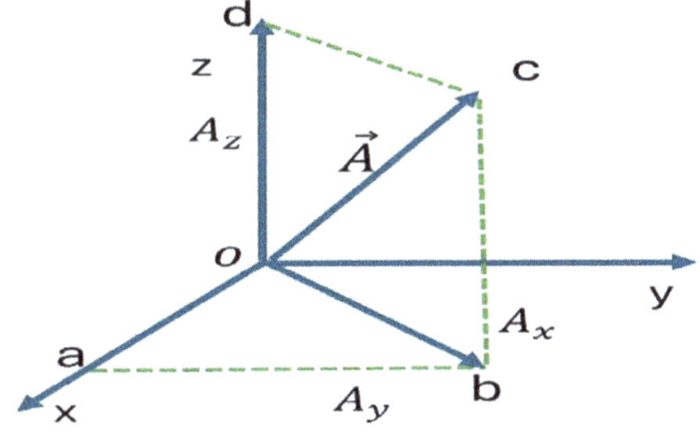

Fig. 1.8 Components of a vector along the three coordinate axes

In the component form, we can represent the vectors as

$$\vec{A} = A_x\hat{x} + A_y\hat{y} + A_z\hat{z} \tag{1.8}$$

$$\vec{B} = B_x\hat{x} + B_y\hat{y} + B_z\hat{z} \tag{1.9}$$

Adding Eqs. (1.8) and (1.9), we get.

$$\vec{A} + \vec{B} = (A_x + B_x)\hat{x} + \left(A_y + B_y\right)\hat{y} + (A_z + B_z)\hat{z} \tag{1.10}$$

The above equation can be written in compact notation as follows:

$$\vec{A} + \vec{B} = \sum_{i=1}^{3}(A_i + B_i)\hat{i} \tag{1.11}$$

Similarly, the dot product of Eqs. (1.8) and (1.9) can be written as follows:

$$\vec{A} \cdot \vec{B} = (A_xB_x) + \left(A_yB_y\right) + (A_zB_z) = \sum_{i=1}^{3}(A_i.B_i) \tag{1.12}$$

In particular,

$$\vec{A} \cdot \vec{A} = A_x^2 + A_y^2 + A_z^2 \tag{1.13}$$

$$\left|\vec{A}\right| = \sqrt{A_x^2 + A_y^2 + A_z^2} \tag{1.14}$$

If we require to find A_x, A_y and A_z components. These components can be obtained as follows:

It is worthwhile that the dot product of any vector $\vec{A}$ with any unit vector is the component of $\vec{A}$ along that direction. Thus, $\hat{x}.\vec{A} = A_x$; $\hat{y}.\vec{A} = A_y$ and $\hat{z}.\vec{A} = A_z$.

The vector product of two vectors $\vec{A}$ and $\vec{B}$ is tedious to calculate by actual multiplication method; however, it can be easily evaluated by expanding the following determinant

$$\vec{A} \times \vec{B} = \begin{vmatrix} \hat{x} & \hat{y} & \hat{z} \\ A_x & A_y & A_z \\ B_x & B_y & B_z \end{vmatrix} \tag{1.15}$$

Example 1.3 Deduce the angle between the face diagonals of unit cube as shown in Fig. 1.9.

Solution: The face diagonals of the cube are

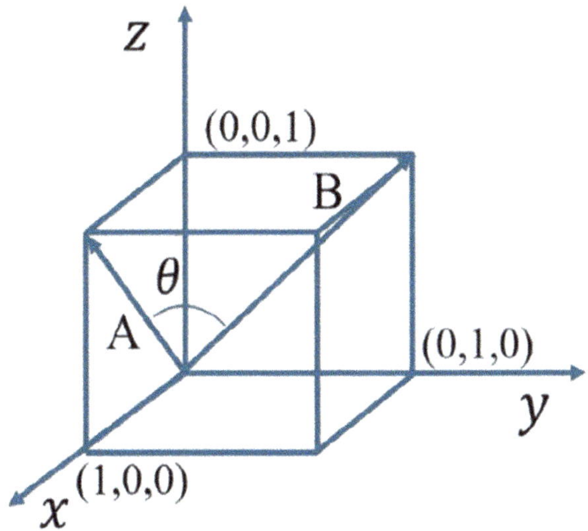

Fig. 1.9 A cube of unit dimensions

$$\vec{A} = \hat{x} + \hat{z} \quad \vec{B} = \hat{y} + \hat{z}$$

$$\vec{A} \cdot \vec{B} = \left|\vec{A}\right|\left|\vec{B}\right|\cos(\theta) = (\hat{x} + \hat{z}) \cdot (\hat{y} + \hat{z})$$

$$\vec{A} \cdot \vec{B} = \hat{z} \cdot \hat{z} = 1$$

$$\left|\vec{A}\right|\left|\vec{B}\right|\cos(\theta) = 1 \tag{1.16}$$

However,

$$\left|\vec{A}\right| = \sqrt{1 + 1} = \sqrt{2} \tag{1.17}$$

Also

$$\left|\vec{B}\right| = \sqrt{1 + 1} = \sqrt{2} \tag{1.18}$$

Substitute Eqs. (1.17) and (1.18) in Eq. (1.15), we get

$$\sqrt{2}\sqrt{2}\cos(\theta) = 1$$

$$\theta = 60^0$$

1.9 Scalar Triple Product

It is written as $\vec{A}.\left(\vec{B} \times \vec{C}\right)$ and is dubbed as the scalar triple product between three vectors $\vec{A}, \vec{B}$ and $\vec{C}$. Geometrically, $|\vec{A}.\left(\vec{B} \times \vec{C}\right)|$ represents the volume of the parallelepiped generated by $\vec{A}, \vec{B}$ and $\vec{C}$, where $\left|\left(\vec{B} \times \vec{C}\right)\right|$ is the area of the base and $|A \cos \theta|$ is the altitude (Fig. 1.10).

$$\vec{B} \times \vec{C} = \left|\vec{B}\right|\left|\vec{C}\right| \sin(\theta)\hat{n}$$

$$\vec{A}.\left(\vec{B} \times \vec{C}\right) = \left|\vec{A}\right|\left|\vec{B}\right|\left|\vec{C}\right| \sin(\theta)\hat{n}.\hat{n}$$

$$\vec{A}.\left(\vec{B} \times \vec{C}\right) = \left|\vec{A}\right|\left|\vec{B}\right|\left|\vec{C}\right| \sin(\theta) \tag{1.19}$$

which represents the volume of the parallelepiped as shown in Fig. 1.10.

1.9.1 Properties of Scalar Triple Product

$$\vec{A}.\left(\vec{B} \times \vec{C}\right) = \vec{B}.\left(\vec{A} \times \vec{C}\right) = \vec{C}.\left(\vec{A} \times \vec{B}\right) \tag{1.20}$$

$$\vec{A}.\left(\vec{C} \times \vec{B}\right) = \vec{B}.\left(\vec{C} \times \vec{A}\right) = \vec{C}.\left(\vec{B} \times \vec{A}\right) \tag{1.21}$$

In the determinant form, we can write scalar triple product as follows:

$$\vec{A}.\left(\vec{B} \times \vec{C}\right) = \begin{vmatrix} A_x & A_y & A_z \\ B_x & B_y & B_z \\ C_x & C_y & C_z \end{vmatrix} \tag{1.22}$$

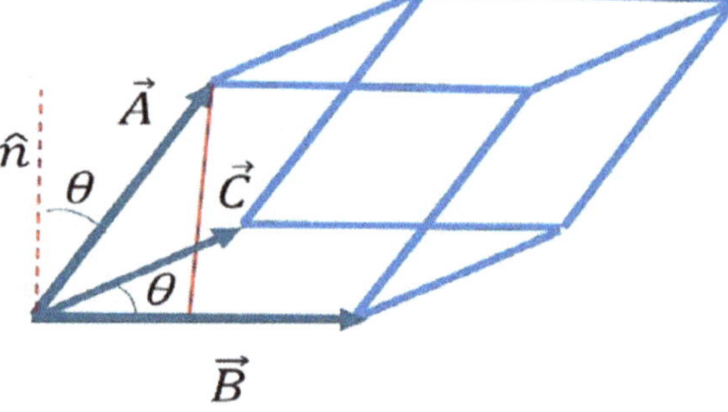

Fig. 1.10 Measurement of the volume using scalar triple product

1.10   Vector Triple Product

This is written as $\vec{A} \times \left(\vec{B} \times \vec{C}\right)$. We can make the observations that $\left(\vec{B} \times \vec{C}\right)$ is perpendicular to the plane containing the vectors $\vec{B}$ and $\vec{C}$. Thus, $\vec{A} \times \left(\vec{B} \times \vec{C}\right)$ is a vector perpendicular to the plane of $\vec{A}$ and $\left(\vec{B} \times \vec{C}\right)$. We are particularly interested in the fact that $\vec{A} \times \left(\vec{B} \times \vec{C}\right)$ is perpendicular to $\left(\vec{B} \times \vec{C}\right)$.

1.10.1   Properties

The vector triple product is a key operation in vector algebra, particularly in physics and engineering. It follows several important properties:

1. **Vector Triple Product Identity**:

    This identity expresses the cross product of a cross product in terms of scalar and vector products, simplifying many vector calculations in mechanics and electromagnetism.

$$\vec{A} \times \left(\vec{B} \times \vec{C}\right) = \vec{B}\left(\vec{A}.\vec{C}\right) - \vec{C}\left(\vec{A}.\vec{B}\right) \tag{1.23}$$

2. **Alternative Form of the Vector Triple Product**:

    This shows an alternative way to rearrange the cross products while maintaining consistency in vector algebra.

$$\vec{A} \times \left(\vec{B} \times \vec{C}\right) = -\vec{C} \times \left(\vec{A} \times \vec{B}\right) = -\vec{A}\left(\vec{B}.\vec{C}\right) + \vec{B}\left(\vec{A}.\vec{C}\right) \tag{1.24}$$

3. **Scalar Quadruple Product**:

    This identity is particularly useful in determining relationships between vectors in space and plays a key role in electrodynamics and mechanics.

$$\left(\vec{A} \times \vec{B}\right).\left(\vec{C} \times \vec{D}\right) = \left(\vec{A}.\vec{C}\right)\left(\vec{B}.\vec{D}\right) - \left(\vec{A}.\vec{D}\right)\left(\vec{B}.\vec{C}\right) \tag{1.25}$$

4. **Cross Product of a Vector Triple Product**:

    This identity extends the triple product to four vectors, often used in advanced physics and engineering calculations.

$$\vec{A} \times \left(\vec{B} \times \left(\vec{C} \times \vec{D}\right)\right) = \vec{B}\left(\vec{A}.\left(\vec{C} \times \vec{D}\right)\right) - \left(\vec{A}.\vec{B}\right)\left(\vec{C} \times \vec{D}\right) \tag{1.26}$$

Example 1.4 Given three vectors $\vec{A} = 2\hat{i}+\hat{j}-3\hat{k}, \vec{B} = -\hat{i}-\hat{j}+4\hat{k}, \vec{C} = 3\hat{i}+2\hat{j}-\hat{k}$. Compute the vector triple product $\vec{A} \times \left(\vec{B} \times \vec{C}\right)$. Using the vector triple product identity $\vec{A} \times \left(\vec{B} \times \vec{C}\right) = \left(\vec{A}.\vec{C}\right)\vec{B} - \left(\vec{A}.\vec{B}\right)\vec{C}$.

Solution:

First, compute dot products

$$\vec{A}.\vec{C} = (2)(3) + (1)(2) + (-3)(-1) = 11$$

$$\vec{A}.\vec{B} = (2)(-1) + (1)(-1) + (-3)(4) = -15$$

Now, apply the triple product rule

$$\vec{A} \times \left(\vec{B} \times \vec{C}\right) = 11\vec{B} + 15\vec{C}$$

Substitute the vectors $\vec{B}$ and $\vec{C}$

$$11\vec{B} = -11\hat{i} - 11\hat{j} + 44\hat{k}$$
$$- 15\vec{C} = -45\hat{i} - 30\hat{j} + 15\hat{k}$$

Thus, $\vec{A} \times \left(\vec{B} \times \vec{C}\right) = 34\hat{i} + 19\hat{j} + 29\hat{k}$.

Example 1.5 A force $\vec{F} = 5\hat{i} + 3\hat{j} - \hat{k}$ is applied to a point located at a position $\vec{r} = 2\hat{i} + \hat{j} + \hat{k}$. The point is moving with a velocity $\vec{v} = -\hat{i} + 2\hat{j} + 3\hat{k}$. Find the torque $\vec{\tau}$ on the point using the vector triple product $\vec{\tau} = \vec{r} \times \left(\vec{v} \times \vec{F}\right)$.

Solution:

First, compute $\vec{v} \times \vec{F}$

$$\vec{v} \times \vec{F} = \begin{vmatrix} \hat{i} & \hat{j} & \hat{k} \\ -1 & 2 & 3 \\ 5 & 3 & -1 \end{vmatrix} = -11\hat{i} + 14\hat{j} - 13\hat{k}$$

Now, compute $\vec{\tau} = \vec{r} \times \left(\vec{v} \times \vec{F}\right) = \begin{vmatrix} \hat{i} & \hat{j} & \hat{k} \\ 2 & 1 & 1 \\ -11 & 14 & -13 \end{vmatrix} = -27\hat{i} + 15\hat{j} + 39\hat{k}$.

Thus, the torque is $\vec{\tau} = -27\hat{i} + 15\hat{j} + 39\hat{k}$.

1.11 Transformation of Vectors

Vectors are defined in coordinate system, and if we change coordinate system, the vector components will be affected. There is a particular geometrical transformation law that governs how the vector components are converted from one frame to another. That is why tensors came into play. Tensors are invariant with respect to coordinate frame.

$$\frac{A_x}{A} = \cos(\theta); \; A_x = A\,\cos(\theta); \; \frac{A_y}{A} = \sin(\theta), A_y = A\,\sin(\theta) \tag{1.27}$$

Let us rotate, the coordinate system by an angle ϕ.
It is obvious from Fig. 1.11 that $\theta' + \phi = \theta$

$$\theta' = \theta - \phi$$
$$A'_x = A\,\cos(\theta - \phi)$$
$$A'_x = A\,\cos(\theta)\cos(\phi) + A\,\sin(\theta)\sin(\phi)$$
$$A'_x = A_x\cos(\phi) + A_y\sin(\phi)$$

Similarly, $A'_y = A\,\sin(\theta - \phi)$

$$A'_y = A\,\sin(\theta)\cos(\phi) - A\,\cos(\theta)\sin(\phi)$$
$$A'_y = A_y\cos(\phi) - A_x\sin(\phi)$$

Thus, the rotation of coordinate system leads to the formation of new components of a vector as follows:

$$A'_x = A_x\cos(\phi) + A_y\sin(\phi), \quad A'_y = A_y\cos(\phi) - A_x\sin(\phi) \tag{1.28}$$

The above conclusion can be expressed in matrix notation as follows:

$$\begin{pmatrix} A'_x \\ A'_y \end{pmatrix} = \begin{pmatrix} \cos(\phi) & \sin(\phi) \\ -\sin(\phi) & \cos(\phi) \end{pmatrix} \begin{pmatrix} A_x \\ A_y \end{pmatrix} \tag{1.29}$$

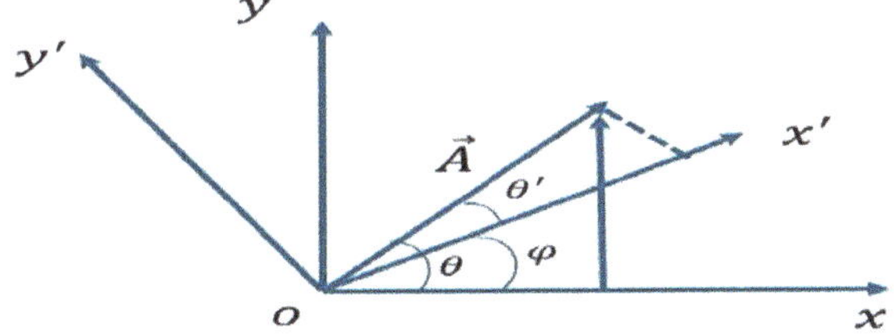

Fig. 1.11 Rotation of a coordinate system

However, the transformation law assumes the following form for rotation about any arbitrary axis in three dimensions

$$\begin{pmatrix} A'_x \\ A'_y \\ A'_z \end{pmatrix} = \begin{pmatrix} R_{xx} & R_{xy} & R_{xz} \\ R_{yx} & R_{yy} & R_{yz} \\ R_{zx} & R_{zy} & R_{zz} \end{pmatrix} \begin{pmatrix} A_x \\ A_y \\ A_z \end{pmatrix} \tag{1.30}$$

Or, more compactly, the above result can be expressed as under

$$A'_i = \sum_{j=1}^{3} R_{ij} A_j \tag{1.31}$$

For any given rotation, we can evaluate the elements of matrix R. Here, we will introduce the concept of a tensor. A tensor of rank (or order) zero is called scalar, and a tensor of rank one is just a vector. In three-dimensional space a scalar has $3^0 = 1$ component and a vector has $3^1 = 3$ components; a second rank tensor has $3^2 = 9$ components; and in general, a tensor of rank n has 3^n components. Moreover, a second rank tensor transforms with two factors of R as follows:

$$T'_{xx} = R_{xx}\left(R_{xx}T_{xx} + R_{xy}T_{xy} + R_{xz}T_{xz}\right) + R_{xy}\left(R_{xx}T_{yx} + R_{xy}T_{yy} + R_{xz}T_{yz}\right)$$
$$+ R_{xz}\left(R_{xx}T_{zx} + R_{xy}T_{zy} + R_{xz}T_{zz}\right)$$

In compact notation, the above result will be written as under

$$T'_{ij} = \sum_{k=1}^{3} \sum_{l=1}^{3} R_{ik} R_{jl} T_{kl} \tag{1.32}$$

Scalar: A rank zero tensor is called a scalar. The various examples are temperature, mass and speed.

Vector: A rank 1 tensor is called a vector. For instance, force, momentum, acceleration, weight, torque and angular momentum. Let us consider a vector as follows:

$$\vec{a} = a_x \hat{x} + a_y \hat{y} + a_z \hat{z}$$

Or, in particular

$$\vec{a} = a_x \hat{e}_1 + a_y \hat{e}_2 + a_z \hat{e}_2$$

In compact form, we can write

$$\vec{a} = \sum_{i=1}^{3} a_i \hat{e}_i \tag{1.33}$$

Example 1.6 A vector $\vec{A} = 3\hat{i} + 4\hat{j}$ is rotated counterclockwise by 45^0 in the *xy*-plane. Find the components of the new vector after rotation.

Solution:

To rotate a vector by an angle θ in 2D, we use the rotation matrix

$$\begin{pmatrix} A'_x \\ A'_y \end{pmatrix} = \begin{pmatrix} \cos\theta & \sin\theta \\ -\sin\theta & \cos\theta \end{pmatrix} \begin{pmatrix} A_x \\ A_y \end{pmatrix}$$

Here, $\theta = 45^0$, $A_x = 3$ and $A_y = 4$. Thus, the rotation matrix becomes

$$\begin{pmatrix} A'_x \\ A'_y \end{pmatrix} = \begin{pmatrix} \frac{1}{\sqrt{2}} & \frac{1}{\sqrt{2}} \\ -\frac{1}{\sqrt{2}} & \frac{1}{\sqrt{2}} \end{pmatrix} \begin{pmatrix} 3 \\ 4 \end{pmatrix}$$

Now, calculate the new components

$$A'_x = 4.94$$
$$A'_y = 0.70$$

Thus, the new components are $\vec{A}' = 4.94\hat{i} + 0.70\hat{j}$.

Example 1.7 A vector $\vec{B} = 2\hat{i} + \hat{j} + \hat{k}$ is rotated by 90^0 about the *z*-axis. Find the new components of the vector after the transformation?

Solution:

$$R_z(90^0) = \begin{pmatrix} \cos 90^0 & -\sin 90^0 & 0 \\ \sin 90^0 & \cos 90^0 & 0 \\ 0 & 0 & 1 \end{pmatrix} = \begin{pmatrix} 0 & -1 & 0 \\ 1 & 0 & 0 \\ 0 & 0 & 1 \end{pmatrix}$$

Now, apply the transformation matrix to $\vec{B}'$

$$\begin{pmatrix} B'_x \\ B'_y \\ B'_z \end{pmatrix} = \begin{pmatrix} 0 & -1 & 0 \\ 1 & 0 & 0 \\ 0 & 0 & 1 \end{pmatrix} \begin{pmatrix} 2 \\ 1 \\ 1 \end{pmatrix}$$

Hence, $B'_x = -1$, $B'_y = 2$ and $B'_z = 1$

$$\vec{B}' = -\hat{i} + 2\hat{j} + \hat{k}$$

1.12 Kronecker Delta $\left(\delta_{ij}\right)$

An isotropic tensor of second rank is called a Kronecker delta. It has, therefore, $3^2 = 9$ components. The one-dimensional Kronecker delta δ_{ij} is defined as

$$\delta_{ij} = \begin{cases} 0, \text{ if } i \neq j \\ 1, \text{ if } i = j \end{cases} \tag{1.34}$$

and

$$\delta_i = \begin{cases} 0, \text{ if } i \neq 0 \\ 1, \text{ if } i = 0 \end{cases} \tag{1.35}$$

It implies that $\delta_{11} = \delta_{22} = \delta_{33} = 1$; $\delta_{12} = \delta_{21} = \delta_{13} = \delta_{31} = \delta_{23} = \delta_{32} = 0$; indicating that δ_{ij} is symmetrical in nature $\left(\delta_{ij} = \delta_{ji}\right)$, and $\delta_{ij}\delta_{jk} = \delta_{ik}$. Now, if j runs from 1 to n, then

$$\sum_{j=1}^{n} \delta_{jj} = \delta_{11} + \delta_{22} + \delta_{33} + \cdots \delta_{nn} = n \tag{1.36}$$

In three-dimensional space $(n = 3)$, we get: $\sum_{j=1}^{3} \delta_{jj} = 3$.

1.13 Scalar Product/Inner Product

As discussed in the previous sections the scalar product of two vectors $\vec{a}$ and $\vec{b}$ is a scalar and is, therefore, defined as follows:

$$\vec{a} = a_x\hat{x} + a_y\hat{y} + a_z\hat{z} = a_i\hat{e}_i \tag{1.37}$$

$$\vec{b} = b_x\hat{x} + b_y\hat{y} + b_z\hat{z} = b_j\hat{e}_j \tag{1.38}$$

$$\vec{a} \cdot \vec{b} = a_i\hat{e}_i \cdot b_j\hat{e}_j = a_ib_j\hat{e}_i \cdot \hat{e}_j \tag{1.39}$$

Since, $\hat{e}_i.\hat{e}_j = \delta_{ij}$, therefore, we obtain from the above equation

$$\vec{a}.\vec{b} = a_ib_j\delta_{ij} \tag{1.40}$$

For $i = j$, we have

$$\vec{a}.\vec{b} = a_i b_i \tag{1.41}$$

$$\vec{a}.\vec{b} = a_1 b_1 + a_2 b_2 + a_3 b_3 \tag{1.42}$$

1.14 Levi–Civita Symbol

It is convenient to introduce the three-dimensional Levi–Civita symbol ϵ_{ijk} which is antisymmetric with respect to all index pairs. It is a third-rank pseudotensor, meaning it behaves as a tensor under proper rotations but changes sign under improper transformations (such as reflections).

The number of components is given by:

$$3^3 = 27$$

The Levi-Civita symbol is defined as:

$$\epsilon_{ijk} = \begin{cases} +1 & \text{for cyclic permutations of } (1, 2, 3) \\ 0 & \text{if any two indices are equal} \\ -1 & \text{acyclic (inverse) permutations of } (1, 3, 2) \end{cases} \tag{1.43}$$

Out of the 27 components:

- 3 components are $+1$

$$\epsilon_{123} = \epsilon_{231} = \epsilon_{312} = +1$$

- 3 components are -1

$$\epsilon_{321} = \epsilon_{213} = \epsilon_{132} = -1$$

- The remaining 21 components are 0, as they involve repeated indices.

1.15 Relation Between ϵ_{ijk} with δ_{ij}

We can find other isotropic tensors from direct products of the two we have or from direct products followed by contraction. It is pertinent to mention here that the direct product of two tensors of ranks n and m is a tensor of rank $n + m$ and that each contraction produces another tensor of rank smaller by 2. If the tensors you multiply are isotropic, the products are also isotropic. By simplifying the products of two Levi–Civita tensors, we can develop a relationship between a Levi–Civita tensor and a Kronecker delta.

$$\epsilon_{ijk}\epsilon_{lmn} = \begin{vmatrix} \delta_{il} & \delta_{im} & \delta_{in} \\ \delta_{jl} & \delta_{jm} & \delta_{jn} \\ \delta_{kl} & \delta_{km} & \delta_{kn} \end{vmatrix} \tag{1.44}$$

On expanding the determinant, we get

$$\epsilon_{ijk}\epsilon_{lmn} = \delta_{il}\left(\delta_{jm}\delta_{kn} - \delta_{jn}\delta_{km}\right) - \delta_{im}\left(\delta_{jl}\delta_{kn} - \delta_{jn}\delta_{kl}\right) + \delta_{in}\left(\delta_{jl}\delta_{km} - \delta_{jm}\delta_{kl}\right)$$

Here summation runs from 1 to 3.

$$\epsilon_{ijk}\epsilon_{lmn} = \delta_{il}\delta_{jm}\delta_{kn} - \delta_{il}\delta_{jn}\delta_{km} - \delta_{im}\delta_{jl}\delta_{kn} + \delta_{im}\delta_{jn}\delta_{kl} + \delta_{in}\delta_{jl}\delta_{km} - \delta_{in}\delta_{jm}\delta_{kl} \tag{1.45}$$

In this, we put $l = i$, and hence the above equation is simplified as follows:

$$\epsilon_{ijk}\epsilon_{imn} = 3\delta_{jm}\delta_{kn} - 3\delta_{jn}\delta_{km} - \delta_{jm}\delta_{kn} + \delta_{km}\delta_{jn} + \delta_{jn}\delta_{km} - \delta_{kn}\delta_{jm} \tag{1.46}$$

Remember that ϵ is zero, unless its three indices are all different. Since the first index is same in ϵ_{ijk} **and** ϵ_{lmn}**,** the product is different from zero only if the other two indices (j, k and m, n) are the same pair in both the epsilons. Thus, Eq. (1.46) attains the following from

$$\epsilon_{ijk}\epsilon_{imn} = \delta_{jm}\delta_{kn} - \delta_{jn}\delta_{km} \tag{1.47}$$

Both sides of Eq. (1.47) are 4th-rank tensors (contracted 6th¯rank tensor on the left) with free indices j, k, m and n. Next, we can see if four indices are same, i.e., if $j = m$. We get from Eq. (1.46)

$$\epsilon_{imk}\epsilon_{imn} = \delta_{mm}\delta_{kn} - \delta_{mn}\delta_{km}$$
$$\epsilon_{imk}\epsilon_{imn} = 3\delta_{kn} - \delta_{kn}$$
$$\epsilon_{imk}\epsilon_{imn} = 2\delta_{kn} \tag{1.48}$$

Further, if all the six indices are same, i.e., if $k = n$

$$\epsilon_{imk}\epsilon_{imk} = 6 \tag{1.49}$$

The familiar formulae in vector analysis can be interpreted in tensor form on using δ_{ij} and ϵ_{ijk}.

These results show that familiar vector identities can be expressed concisely using the Kronecker delta and the Levi-Civita symbol in tensor notation.

1.16   Vector Product in Tensor Form

The cross product of two vectors $\vec{a} \times \vec{b}$ in three-dimensional space can be expressed using the Levi-Civita symbol ϵ_{ijk}. This formulation provides a tensorial representation of the cross product. Using the determinant definition of the cross product:

The components of the cross product of two vectors can be written as

$$\vec{a} \times \vec{b} = \begin{vmatrix} \hat{e}_1 & \hat{e}_2 & \hat{e}_3 \\ a_x & a_y & a_z \\ b_x & b_y & b_z \end{vmatrix}$$

$$\vec{a} \times \vec{b} = \hat{e}_1\left(a_y b_z - a_z b_y\right) - \hat{e}_2(a_x b_z - a_z b_x) + \hat{e}_3\left(a_x b_y - a_y b_x\right)$$

This result shows that each component of the cross product is a linear combination of the components of $\vec{a}$ and $\vec{b}$, with alternating signs. From tensor notation, the ith component of the cross product can be written using the Levi-Civita symbol.

$$\left(\vec{a} \times \vec{b}\right)_i = \hat{e}_i \epsilon_{ijk} a_j b_k$$

$$\left(\vec{a} \times \vec{b}\right)_i = \hat{e}_1 (\epsilon_{123} a_2 b_3 + \epsilon_{132} a_3 b_2)$$

$$\left(\vec{a} \times \vec{b}\right)_i = \hat{e}_1 (a_2 b_3 - a_3 b_2) \tag{1.50}$$

Or,

$$\left(\vec{a} \times \vec{b}\right)_i = \hat{e}_1\left(a_y b_z - a_z b_y\right)$$

Similarly, we can write

$$\left(\vec{a} \times \vec{b}\right)_j = \hat{e}_j \epsilon_{jki} a_k b_i$$

$$\left(\vec{a} \times \vec{b}\right)_j = \hat{e}_2 (\epsilon_{231} a_3 b_1 + \epsilon_{213} a_1 b_3)$$

$$\left(\vec{a} \times \vec{b}\right)_j = \hat{e}_2 (a_3 b_1 - a_1 b_3)$$

$$\left(\vec{a} \times \vec{b}\right)_j = -\hat{e}_2 (a_x b_z - a_z b_x) \tag{1.51}$$

Proceeding in the same manner, we can, therefore, write

$$\left(\vec{a} \times \vec{b}\right)_k = \hat{e}_k \epsilon_{kij} a_i b_j = -\hat{e}_3\left(a_x b_y - a_y b_x\right) \tag{1.52}$$

From the above discussion it is evident that we can express the components of a vector product of two vectors in a simplified form while using the concept of tensors.

1.17 Scalar Triple Product Using Tensors

The scalar product of three vectors is a scalar quantity. We can evaluate it as follows:
The scalar triple product of three vectors $\vec{a}$, $\vec{b}$ and $\vec{c}$ is defined as:

$$\vec{a}.(\vec{b}\times\vec{c})$$

which results in a scalar quantity. This product has significant geometric and algebraic interpretations, making it a fundamental concept in vector and tensor analysis. Using the Einstein summation convention, the scalar triple product can be expressed as:

$$\vec{a}.(\vec{b}\times\vec{c}) = \vec{a}_i.(\vec{b}\times\vec{c})_i$$

Expanding the cross product in index notation using the Levi-Civita symbol ϵ_{ijk}

$$\vec{a}.(\vec{b}\times\vec{c}) = a_i\epsilon_{ijk}b_jc_k$$

Rearranging the indices, we obtain:

$$\vec{a}.(\vec{b}\times\vec{c}) = \epsilon_{ijk}a_ib_jc_k$$

By renaming the dummy indices (which does not change the summation result), we can rewrite:

$$\vec{a} \cdot \left(\vec{b} \times \vec{c}\right) = a_i \cdot \left(\vec{b} \times \vec{c}\right)_i$$
$$\vec{a} \cdot \left(\vec{b} \times \vec{c}\right) = a_i\epsilon_{ijk}b_jc_k$$
$$\vec{a} \cdot \left(\vec{b} \times \vec{c}\right) = \epsilon_{ijk}a_ib_jc_k$$
$$\vec{a} \cdot \left(\vec{b} \times \vec{c}\right) = \epsilon_{kij}a_ib_jc_k$$
$$\vec{a} \cdot \left(\vec{b} \times \vec{c}\right) = c_k\epsilon_{kij}a_ib_j$$
$$\vec{a} \cdot \left(\vec{b} \times \vec{c}\right) = c_k\left(\vec{a} \times \vec{b}\right)_k$$
$$\vec{a} \cdot \left(\vec{b} \times \vec{c}\right) = \vec{c} \cdot \left(\vec{a} \times \vec{b}\right) \tag{1.53}$$

Proceeding in the same manner, we can show that

$$\epsilon_{ijk}a_ib_jc_k = b_j\epsilon_{jki}c_ka_i = b_j(\vec{c}\times\vec{a})_j = \vec{b}\cdot(\vec{c}\times\vec{a}) \tag{1.54}$$

Hence, we conclude from the above discussion that

$$\vec{a}\cdot\left(\vec{b}\times\vec{c}\right) = \vec{c}\cdot\left(\vec{a}\times\vec{b}\right) = \vec{b}\cdot(\vec{c}\times\vec{a}) \tag{1.55}$$

1.18 Vectors Triple Product Using Tensors

We can make use of formulae developed in the preceding sections so as to simplify the vector triple product as under

$$\vec{a}\times\left(\vec{b}\times\vec{c}\right) = \vec{b}(\vec{a}\cdot\vec{c}) - \vec{c}\left(\vec{a}\cdot\vec{b}\right)$$

$$\vec{a}\times\left(\vec{b}\times\vec{c}\right) = \left(\vec{a}\times\left(\vec{b}\times\vec{c}\right)\right)_i = \epsilon_{ijk}a_j\left(\vec{b}\times\vec{c}\right)_k$$

$$= \epsilon_{ijk}a_j\epsilon_{klm}b_lc_m$$

$$= \epsilon_{ijk}\epsilon_{klm}a_jb_lc_m$$

Since we know that if two indices of Levi–Civita tensors are same, it can be manifested with regard to Kronecker delta as follows:

$$\left(\vec{a}\times\left(\vec{b}\times\vec{c}\right)\right)_i = \left(\delta_{il}\delta_{jm} - \delta_{im}\delta_{jl}\right)a_jb_lc_m$$

$$= \delta_{il}\delta_{jm}a_jb_lc_m - \delta_{im}\delta_{jl}a_jb_lc_m$$

$$= \left(\delta_{il}b_l\right)\left(\delta_{jm}a_jc_m\right) - \left(\delta_{im}c_m\right)\left(\delta_{jl}a_jb_l\right)$$

If, we put, $i = l, j = m$, and $j = l$ and $m = i$ respectively in the first and second terms on the RHS of above expression,

$$\left(\vec{a}\times\left(\vec{b}\times\vec{c}\right)\right)_i = \left(\delta_{ii}b_i\right)\left(\delta_{jj}a_jc_j\right) - \left(\delta_{ii}c_i\right)\left(\delta_{jj}a_jb_j\right)$$

As we consider only one component $\delta_{ii} = \delta_{jj} = 1$

$$\left(\vec{a}\times\left(\vec{b}\times\vec{c}\right)\right)_i = \left(b_i\right)\left(a_jc_j\right) - \left(c_i\right)\left(a_jb_j\right)$$

Hence

$$\left(\vec{a}\times\left(\vec{b}\times\vec{c}\right)\right)_i = \left(\vec{b}(\vec{a}.\vec{c}) - \vec{c}\left(\vec{a}.\vec{b}\right)\right)_i \tag{1.56}$$

1.19   The Del Operator

It is to be noted that we recognize $\vec{\nabla}$ as if it is vector. Here, we can similarly treat it as a first rank tensor, always remembering that it is also a differential operator. We can, therefore, establish the following formula for curl of $\left(\vec{V} \times \vec{W}\right)$ by making use of tensors

$$\vec{\nabla} \times \left(\vec{V} \times \vec{W}\right) = \left(\vec{W} \cdot \vec{\nabla}\right)\vec{V} + \left(\vec{\nabla} \cdot \vec{W}\right)\vec{V} - \left(\vec{\nabla} \cdot \vec{V}\right)\vec{W} - \left(\vec{V} \cdot \vec{\nabla}\right)\vec{W}$$

We know that the vector triple product in tensor form is written as follows:

$$\left(\vec{\nabla} \times \left(\vec{V} \times \vec{W}\right)\right)_i = \epsilon_{ijk} \vec{\nabla}_j \epsilon_{klm} V_l W_m$$

$$\left(\vec{\nabla} \times \left(\vec{V} \times \vec{W}\right)\right)_i = \epsilon_{ijk} \epsilon_{klm} \vec{\nabla}_j (V_l W_m)$$

$$\left(\vec{\nabla} \times \left(\vec{V} \times \vec{W}\right)\right)_i = \epsilon_{ijk} \epsilon_{klm} \left[\left(\vec{\nabla}_j V_l\right) W_m + V_l \left(\vec{\nabla}_j W_m\right)\right]$$

$$= \left(\delta_{il}\delta_{jm} - \delta_{im}\delta_{jl}\right)\left[\left(\vec{\nabla}_j V_l\right) W_m + V_l \left(\vec{\nabla}_j W_m\right)\right]$$

$$\left(\vec{\nabla} \times \left(\vec{V} \times \vec{W}\right)\right)_i = \delta_{il}\delta_{jm}\left(\vec{\nabla}_j V_l\right) W_m$$

$$- \delta_{lm}\delta_{jl}\left(\vec{\nabla}_j V_l\right) W_m + \delta_{il}\delta_{jm} V_l \left(\vec{\nabla}_j W_m\right) - \delta_{im}\delta_{jl} V_l \left(\vec{\nabla}_j W_m\right)$$

Let in the above expression $i = l, j = m$ (in first term), $i = m, j = l$ (in 2nd term), $i = l, j = m$ (in 3rd term), $i = m, j = l$ (in 4th term).

$$\left(\vec{\nabla} \times \left(\vec{V} \times \vec{W}\right)\right)_i = \left(\vec{\nabla}_m V_i\right) W_m - \left(\vec{\nabla}_j V_j\right) W_i + V_i \left(\vec{\nabla}_m W_m\right) - V_j \left(\vec{\nabla}_j W_i\right)$$

$$\left(\vec{\nabla} \times \left(\vec{V} \times \vec{W}\right)\right)_i = W_m \left(\vec{\nabla}_m V_i\right) - \left(\vec{\nabla}_j V_j\right) W_i + V_i \left(\vec{\nabla}_m W_m\right) - \left(V_j \vec{\nabla}_j\right) W_i$$

Hence, we can write

$$\vec{\nabla} \times \left(\vec{V} \times \vec{W}\right) = \left(\vec{W} \cdot \vec{\nabla}\right)\vec{V} + \left(\vec{\nabla} \cdot \vec{W}\right)\vec{V} - \left(\vec{\nabla} \cdot \vec{V}\right)\vec{W} - \left(\vec{V} \cdot \vec{\nabla}\right)\vec{W} \quad (1.57)$$

Example 1.8 Simplify the expression $\epsilon_{ijk}\epsilon_{lmn}$ using Kronecker deltas and show that

$$\epsilon_{ijk}\epsilon_{lmn} = \delta_{il}\left(\delta_{jm}\delta_{kn} - \delta_{jn}\delta_{km}\right) - \delta_{im}\left(\delta_{jl}\delta_{kn} - \delta_{jn}\delta_{kl}\right) + \delta_{in}\left(\delta_{jl}\delta_{km} - \delta_{jm}\delta_{kl}\right)$$

Solution:

The product of two Levi–Civita symbols can be expressed as a determinant involving Kronecker deltas, as stated in the problem. The general form is:

$$\epsilon_{ijk}\epsilon\epsilon_{lmn} = \begin{vmatrix} \delta_{il} & \delta_{im} & \delta_{in} \\ \delta_{jl} & \delta_{jm} & \delta_{jn} \\ \delta_{kl} & \delta_{km} & \delta_{kn} \end{vmatrix}$$

On expansion of the determinant, we have

$$\epsilon_{ijk}\epsilon_{lmn} = \delta_{il}\left(\delta_{jm}\delta_{kn} - \delta_{jn}\delta_{km}\right) - \delta_{im}\left(\delta_{jl}\delta_{kn} - \delta_{jn}\delta_{kl}\right) + \delta_{in}\left(\delta_{jl}\delta_{km} - \delta_{jm}\delta_{kl}\right) \quad (1.58)$$

This equation represents the product of two Levi–Civita tensors expressed as a combination of Kronecker deltas as required.

Example 1.9 Use the result $\epsilon_{ijk}\epsilon_{imn} = \delta_{jm}\delta_{kn} - \delta_{jn}\delta_{km}$ to calculate the contracted tensor $\epsilon_{imk}\epsilon_{imn}$.

Solution:

Since we know that

$$\epsilon_{ijk}\epsilon_{imn} = \delta_{jm}\delta_{kn} - \delta_{jn}\delta_{km}$$

Setting $j = m$, the above expression becomes

$$\epsilon_{imk}\epsilon_{imn} = \delta_{mm}\delta_{kn} - \delta_{mn}\delta_{km}$$

First, compute δ_{mm} which is the summation over the repeated index m

$$\delta_{mm} = 3$$

Thus, the equation becomes

$$\epsilon_{imk}\epsilon_{imn} = 3\delta_{kn} - \delta_{mn}\delta_{km} \quad (1.59)$$

Example 1.10 Use the result $\epsilon_{ijk}\epsilon_{imn} = \delta_{jm}\delta_{kn} - \delta_{jn}\delta_{km}$ to calculate $\epsilon_{imk}\epsilon_{imk}$ (i.e., when all the six indices are same).

Solution:

Using the same equation

$$\epsilon_{imk}\epsilon_{imk} = \delta_{mm}\delta_{kk} - \delta_{mk}\delta_{mk}$$

Since in three dimensions ($n = 3$) $\delta_{mm} = 3$, $\delta_{kk} = 3$ and $\delta_{mk}\delta_{mk} = 3$

$$\epsilon_{imk}\epsilon_{imk} = 6 \quad (1.60)$$

This is a familiar result in vector analysis, where the fully contracted Levi–Civita tensor gives 6, which is the number of independent non-zero components of the Levi–Civita symbol in 3D three dimensions.

Example 1.11 Given three vectors $\vec{a}, \vec{b}$ and $\vec{c}$, verify that the scalar triple product $\vec{a} \cdot \left(\vec{b} \times \vec{c} \right)$ can be expressed using the Levi–Civita symbol as $\vec{a} \cdot \left(\vec{b} \times \vec{c} \right) = \epsilon_{ijk} a_i b_j c_k$.

Solution:

We can write the scalar triple product as follows:

$$\vec{a} \cdot \left(\vec{b} \times \vec{c} \right) = a_i (b \times c)_i$$

Using the definition of the cross product in index notation:

$$(b \times c)_i = \epsilon_{ijk} b_j c_k$$

Substitute this into the expression for the scalar triple product

$$\vec{a} \cdot \left(\vec{b} \times \vec{c} \right) = a_i \epsilon_{ijk} b_j c_k$$

Or

$$\vec{a} \cdot \left(\vec{b} \times \vec{c} \right) = \epsilon_{ijk} a_i b_j c_k \tag{1.61}$$

Thus, we have verified that the scalar triple product can be expressed in terms of the Levi-Civita symbol.

1.20   Differential Calculus

Let $f(x)$ be any function. The change in the function with respect to the change in its domain or variable x is called derivative of a function. Physically, the derivative is interpreted as the slope of a function at any arbitrary point in its domain (Fig. 1.12).

$$\frac{\Delta f(x)}{\Delta x} = \frac{f(x_2) - f(x_1)}{x_2 - x_1}$$

If we apply the limit on both sides, we get

$$\lim_{\Delta x \to 0} \frac{\Delta f(x)}{\Delta x} = \lim_{\Delta x \to 0} \frac{f(x_2) - f(x_1)}{x_2 - x_1}$$

$$\frac{df}{dx} = \lim_{\Delta x \to 0} \frac{f(x_2) - f(x_1)}{x_2 - x_1} \tag{1.62}$$

This is called as derivative or slope or rise over run of a function. $\frac{df}{dx}$ is called as spatial derivative; on the other hand, $\frac{df}{dt}$ is called as time derivative or temporal derivative.

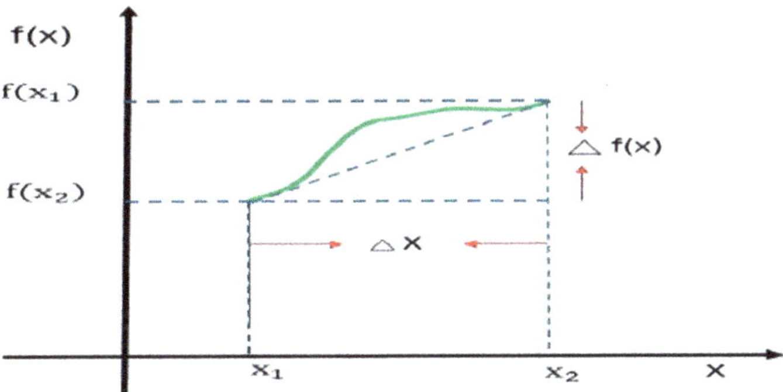

Fig. 1.12 Graphical representation of differentiation of a function

$$df = \left(\frac{df}{dx}\right)dx \tag{1.63}$$

1.21 The Gradient

In order to provide more motivation for the vector nature of partial derivatives, we will introduce the total variation of a function. Let us assume that T be a function that depends on the spatial coordinates x, y and z. Therefore, its total variation is given by

$$dT = \left(\frac{\partial T}{\partial x}\right)dx + \left(\frac{\partial T}{\partial y}\right)dy + \left(\frac{\partial T}{\partial z}\right)dz$$

$$dT = \left(\left(\frac{\partial T}{\partial x}\right)\hat{x} + \left(\frac{\partial T}{\partial y}\right)\hat{y} + \left(\frac{\partial T}{\partial z}\right)\hat{z}\right) \cdot (dx\hat{x} + dy\hat{y} + dz\hat{z})$$

Only the inner components multiply that is why dot product is also called as inner product.

$$dT = \left(\left(\frac{\partial}{\partial x}\right)\hat{x} + \left(\frac{\partial}{\partial y}\right)\hat{y} + \left(\frac{\partial}{\partial z}\right)\hat{z}\right) T \cdot (dx\hat{x} + dy\hat{y} + dz\hat{z})$$

The first term, i.e., $\left(\left(\frac{\partial}{\partial x}\right)\hat{x} + \left(\frac{\partial}{\partial y}\right)\hat{y} + \left(\frac{\partial}{\partial z}\right)\hat{z}\right)$, is defined as spatial derivative called del operator or nabla, represented by $\vec{\nabla}$. It is pertinent to mention here that the del operator is meaningless unless it is operated upon some function. Furthermore, it is not a vector in usual sense.

$$dT = \vec{\nabla}T \cdot d\vec{l} \tag{1.64}$$

$$dT = \left|\vec{\nabla}T\right|\left|d\vec{l}\right|\cos(\theta) \tag{1.65}$$

$\vec{\nabla}T$ represents the spatial derivative of a scalar function T in a specific direction, or it gives slope of a function in a particular direction. It is a vector quantity having three components.

Consider heater the room at corner of a room. There is a particular direction along which there occurs a maximum change. The maximum change in T evidentially occurs in the same direction as that of $\vec{\nabla}T$. Therefore, the gradient $\vec{\nabla}T$ points in the direction of maximum increase of the function T. Moreover, $|\vec{\nabla}T|$ gives the slope along this maximal direction.

Example 1.12 Given the field $\phi(x, y, z) = x^2 + y^2 + z^2$. Calculate the $\vec{\nabla}\phi$.
    Solution:

$$\vec{\nabla}\phi = \left(\frac{\partial\phi}{\partial x}\right)\hat{i} + \left(\frac{\partial\phi}{\partial y}\right)\hat{j} + \left(\frac{\partial\phi}{\partial z}\right)\hat{k}$$

$$\vec{\nabla}\phi = 2x\hat{i} + 2y\hat{j} + 2z\hat{k}$$

Example 1.13 Deduce the gradient of $r = \sqrt{x^2 + y^2 + z^2}$.
    Solution:
Since we are given that (Fig. 1.13).

$$r = \sqrt{x^2 + y^2 + z^2}$$

$$\vec{\nabla}r = \vec{\nabla}\left(\sqrt{x^2 + y^2 + z^2}\right)$$

$$\vec{\nabla}r = \left(\frac{\partial r}{\partial x}\right)\hat{x} + \left(\frac{\partial r}{\partial y}\right)\hat{y} + \left(\frac{\partial r}{\partial z}\right)\hat{z}$$

$$\vec{\nabla}r = \frac{1}{2}\frac{2x}{\sqrt{x^2 + y^2 + z^2}}\hat{x} + \frac{1}{2}\frac{2y}{\sqrt{x^2 + y^2 + z^2}}\hat{y}$$

$$+ \frac{1}{2}\frac{2z}{\sqrt{x^2 + y^2 + z^2}}\hat{z}$$

$$\vec{\nabla}r = \frac{x\hat{x} + y\hat{y} + z\hat{z}}{\sqrt{x^2 + y^2 + z^2}} = \frac{\vec{r}}{r} = \hat{r} \tag{1.66}$$

Physically it can be interpreted as the distance from the origin increases abruptly in the radial direction and its rate of increase in that direction is unity.

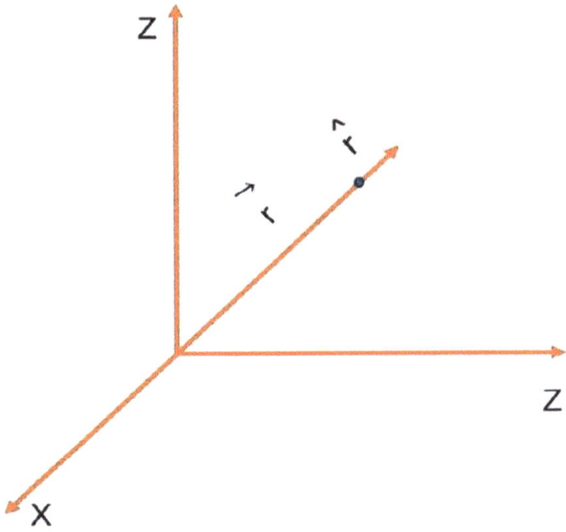

Fig. 1.13 Gradient of radial distance in Cartesian coordinates

1.22 The Divergence

It gives the spread of a vector quantity. The divergence of a vector quantity is a scalar. Consider the vector function

$$\vec{V} = V_x\hat{x} + V_y\hat{y} + V_z\hat{z} \tag{1.67}$$

The divergence of above vector function is defined as follows:

$$\vec{\nabla}.\vec{V} = \left(\left(\frac{\partial}{\partial x}\right)\hat{x} + \left(\frac{\partial}{\partial y}\right)\hat{y} + \left(\frac{\partial}{\partial z}\right)\hat{z}\right).(V_x\hat{x} + V_y\hat{y} + V_z\hat{z})$$

As we know in dot product, only internal components multiply.

$$\vec{\nabla}.\vec{V} = \left(\frac{\partial V_x}{\partial x}\right) + \left(\frac{\partial V_y}{\partial y}\right) + \left(\frac{\partial V_z}{\partial z}\right) \tag{1.68}$$

We will now investigate the meaning and use of divergence in physical applications. Consider a region in which water is flowing. We can imagine by drawing at every point a vector $\vec{V}$ equal to the velocity of the water at that point. The vector function $\vec{V}$ then represents a vector field. The curves tangent to $\vec{V}$ are called streamlines. We could in the same way discuss the flow of a gas, of heat, of electricity, or of particles (say from a radioactive source). We can show that if $\vec{V}$ represents the velocity of flow of any of these things, then div $\vec{V}$ is related to the amount of the substance which flows out of a given volume. This could be different from zero either because of a change in density (more air flows out than in as a room is heated) or

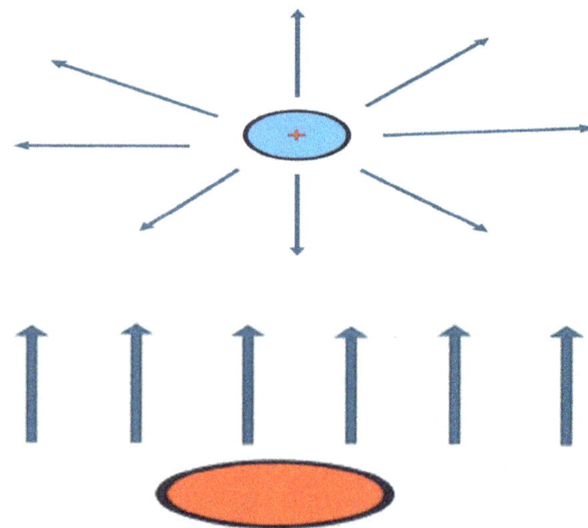

Fig. 1.14 Divergence of a vector field

Fig. 1.15 Zero divergence in a uniform field

because there is a source or sink in the volume (alpha particles flow out of but not into a box containing an alpha-radioactive source). Exactly the same mathematics applies to the electric and magnetic fields where $\vec{V}$ is replaced by $\vec{E}$ or $\vec{B}$ and the quantity corresponding to outflow of a material substance is called flux (Figs. 1.14 and 1.15).

Figure 1.14 illustrates a diverging vector field, where vectors radiate outward from a central point. This represents a source, such as water flowing out of a faucet or air escaping from a balloon. The divergence at these points is positive since the fluid expands outward.

For example, in the case of an expanding gas, if more gas molecules leave a given volume than enter, the divergence is positive, indicating net outflow. Mathematically, this corresponds to:

$$\vec{\nabla}.\vec{V} > 0$$

This also applies to electric fields, where a positive divergence of $\vec{E}$ corresponds to the presence of a positive charge, which acts as a source of the field.

Figure 1.15 likely depicts a converging vector field, where vectors point inward toward a central point. This represents a sink, such as water being drained into a hole or air being sucked into a vacuum pump. The divergence at these points is negative, indicating net inflow of the substance.

For example, if a room is being cooled and air molecules are removed, then $\vec{\nabla}.\vec{V} < 0$ representing net inflow. Similarly, in an electric field, a negative divergence of $\vec{E}$ corresponds to the presence of a negative charge, which acts as a sink of the field.

Example 1.14 Given a temperature distribution in space $T(x, y, z) = x^2 + 4y^2 + 9z^2$, find the gradient $\vec{\nabla}T$ and interpret its physical meaning.

Solution:

The gradient operator of the temperature is written as follows:

$$\vec{\nabla}T = \left(\frac{\partial T}{\partial x}\right)\hat{i} + \left(\frac{\partial T}{\partial y}\right)\hat{j} + \left(\frac{\partial T}{\partial z}\right)\hat{k} \tag{1.69}$$

In the present case, we have $\vec{\nabla}T = 2x\hat{i} + 8y\hat{j} + 18z\hat{k}$.

Physical Interpretation:

- The gradient $\vec{\nabla}T$ points in the direction of the maximum rate of increase of temperature in space.
- The magnitude of the gradient represents how fast the temperature is increasing in that direction.
- At any point, the vector $\vec{\nabla}T$ gives both the direction and the magnitude of the greatest rate of change in temperature.

Example 1.15 Find the rate of change of the scalar field $\phi(x, y, z) = 2x^2 + y^2 - z^2$ at the point $(1, -1, 2)$ along the direction of the vector $\vec{v} = \hat{i} - 2\hat{j} + 2\hat{k}$.

Solution:

The directional derivative of a scalar field ϕ in the direction of a vector $\vec{v}$ is illustrated as under

$$\vec{\nabla}\phi.\hat{v}$$
$$\vec{\nabla}\phi = 4x\hat{i} + 2y\hat{j} - 2z\hat{k}$$

At the point $(1, -1, 2)$, the gradient becomes:

$$\vec{\nabla}\phi = 4\hat{i} - 2\hat{j} - 4\hat{k}$$

The magnitude of vector $\vec{v}$ is $|\vec{v}| = 3$. The unit vector $\hat{v}$ is given by

$$\hat{v} = \frac{1}{3}\left(\hat{i} - 2\hat{j} + 2\hat{k}\right)$$

Therefore, the directional derivative is given by

$$\vec{\nabla}\phi.\hat{v} = 0$$

Example 1.16 If $\vec{V}_a = x\hat{x} + y\hat{y} + z\hat{z}$; $\vec{V}_b = \hat{z}$; $\vec{V}_c = z\hat{z}$. Evaluate the divergence in each case

Solution: $\vec{\nabla}.\vec{V}_a = \left(\frac{\partial x}{\partial x}\right) + \left(\frac{\partial y}{\partial y}\right) + \left(\frac{\partial z}{\partial z}\right) = 3.$

As expected, this function possesses positive divergence

$$\vec{\nabla} \cdot \vec{V}_b = 0$$

Exactly, it was anticipated

$$\vec{\nabla} \cdot \vec{V}_c = 1$$

$$\vec{\nabla} \cdot \vec{V}_b = 0$$

It means inward flux is equal to outward flux.

$$\vec{\nabla} \cdot \vec{V}_c = 1$$

These results have far-reaching consequences in electrodynamics. In electrodynamics, we have to deal with electric and magnetic fields. We are mainly concerned to evaluate the divergence of these fields in electrodynamics. Hence, these results could serve as a cornerstone for electrodynamics.

1.23 The Curl

The curl of any vector function $\vec{V}$ is constructed as follows:

$$\vec{\nabla} \times \vec{V} = \begin{vmatrix} \hat{x} & \hat{y} & \hat{z} \\ \frac{\partial}{\partial x} & \frac{\partial}{\partial y} & \frac{\partial}{\partial z} \\ V_x & V_y & V_z \end{vmatrix}$$

$$\vec{\nabla} \times \vec{V} = \hat{x}\left(\frac{\partial V_z}{\partial y} - \frac{\partial V_y}{\partial z}\right) - \hat{y}\left(\frac{\partial V_z}{\partial x} - \frac{\partial V_x}{\partial z}\right) + \hat{z}\left(\frac{\partial V_y}{\partial x} - \frac{\partial V_x}{\partial y}\right) \qquad (1.70)$$

It is worthwhile to note that the curl of any vector function produces a vector. Geometrically, it measures how much a vector function curls around a particular point.

Example 1.17 Let $\vec{V}_a = -y\hat{x} + x\hat{y}$ be the function sketched in Fig. 1.16. Calculate its curl

Solution:
Since $\vec{V}_a = -y\hat{x} + x\hat{y}$ is the given vector function

Fig. 1.16 Representation of curl in a vector field

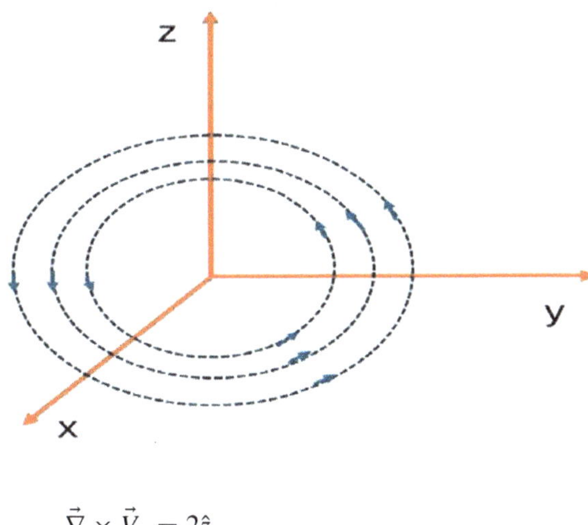

$$\vec{\nabla} \times \vec{V}_a = 2\hat{z}$$

This indicates that the function under consideration has a substantial curl that points in the z-direction.

Example 1.18 Let $\vec{V}_b = x\hat{y}$ be the function sketched in Fig. 1.17. Calculate its curl.
   Solution: Since the given function is $\vec{V}_b = x\hat{y}$

$$\vec{\nabla} \times \vec{V}_b = \hat{z}$$

From the above result, it is evident that the function under consideration has some substantial curl which points in the z-direction.

Fig. 1.17 Vector field representation with curl

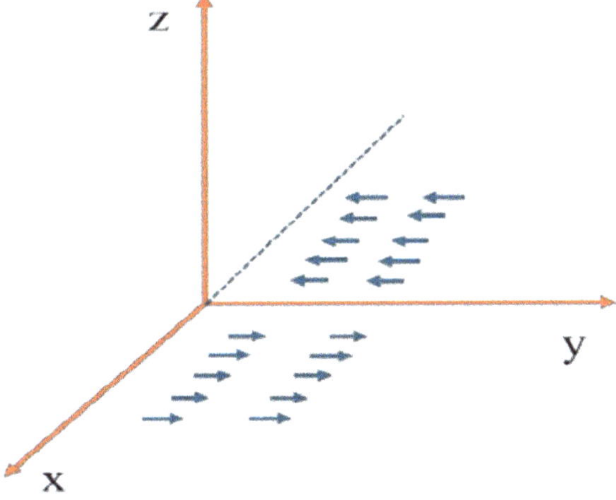

1.24 Second-Order Derivatives

The gradient, the divergence and the curl are the only first derivatives. Further, if we operate $\vec{\nabla}$ twice, we can, therefore, construct five species of second derivatives as follows.

1. **Divergence of gradient**: $\vec{\nabla}.(\vec{\nabla}T)$, This is just Laplacian of T and is, therefore, a scalar quantity.
2. **Curl of gradient**: $\vec{\nabla} \times \left(\vec{\nabla}T\right)$, The curl of gradient is always zero, i.e., $\vec{\nabla} \times \left(\vec{\nabla}T\right) = 0$. However, $\vec{\nabla}$ being an operator and it does not multiply in the usual sense. Its proof hinges on the notion of the equality of cross derivatives.
3. **Gradient of divergence**: $\vec{\nabla}(\vec{\nabla}.\vec{V})$, This quantity is different from Laplacian and often occurs in physical application.
4. **Divergence of curl**: $\vec{\nabla}. (\vec{\nabla} \times \vec{V})$, The divergence of curl is also zero, i.e., $\vec{\nabla}.\left(\vec{\nabla} \times \vec{V}\right) = 0$
5. **Curl of curl**: $\vec{\nabla} \times (\vec{\nabla} \times \vec{V})$, The curl of curl gives nothing new.

$$\vec{\nabla}.\left(\vec{\nabla}T\right) \neq 0$$

$$\vec{\nabla} \times \left(\vec{\nabla}T\right) = 0$$

$$\vec{\nabla}\left(\vec{\nabla}.\vec{V}\right) \neq 0$$

$$\vec{\nabla}.\left(\vec{\nabla} \times \vec{V}\right) = 0$$

$$\vec{\nabla} \times \left(\vec{\nabla} \times \vec{V}\right) \neq 0$$

From the above discussion, it is evident that there are just two second-order derivatives, viz. Laplacian and gradient of divergence. The Laplacian is of prime importance, and the gradient of divergence is seldom encountered in physical problems. However, we could also work out third-order derivatives but for practical purposes only second-order derivatives suffice.

Example 1.19 Given the vector fields $\vec{V} = x\hat{i} + y\hat{j} + z\hat{k}$ and $\vec{W} = -y\hat{i} + x\hat{j} + 2z\hat{k}$. Calculate $\vec{\nabla} \times \left(\vec{V} \times \vec{W}\right)$.

Solution:

We know that

$$\vec{\nabla} \times \left(\vec{V} \times \vec{W}\right) = \left(\vec{W} \cdot \vec{\nabla}\right)\vec{V} + \left(\vec{\nabla} \cdot \vec{W}\right)\vec{V} - \left(\vec{\nabla} \cdot \vec{V}\right)\vec{W} - \left(\vec{V} \cdot \vec{\nabla}\right)\vec{W}$$

$$\left(\vec{W} \cdot \vec{\nabla}\right)\vec{V} = -y\hat{i} + x\hat{j} + 2z\hat{k}, \left(\vec{\nabla} \cdot \vec{W}\right)\vec{V} = 2x\hat{i} + 2y\hat{j} + 2z\hat{k},$$

$$\left(\vec{\nabla} \cdot \vec{V}\right)\vec{W} = -3y\hat{i} + 3x\hat{j} + 6z\hat{k} \text{ and } \left(\vec{V} \cdot \vec{\nabla}\right)\vec{W} = -y\hat{i} + x\hat{j} + 2z\hat{k}$$

Hence

$$\vec{\nabla} \times \left(\vec{V} \times \vec{W}\right) = (2x + 3y)\hat{i} + (2y - 3x)\hat{j} - 4z\hat{k}$$

1.25 Integral Calculus

The integrals which we often encounter in electrodynamics are line (or path) integrals, surface (or flux) integrals and volume integrals.

1.25.1 Line Integrals

It can be represented as $\int_a^b \vec{V} \cdot d\vec{l}$, where $\vec{V}$ being some vector function and $d\vec{l}$ is an infinitesimal displacement vector. However, if the path under consideration forms a closed loop, then we mark a circle sign on the integral, i.e., $\oint \vec{V} \cdot d\vec{l}$. The quantity $\vec{V} \cdot d\vec{l}$ represents area under the curve. For instance, work performed by a constant force; i.e., $W = \int_a^b \vec{F} \cdot d\vec{l}$, is the most familiar example of the line integral that we often come across in physics (Fig. 1.18).

The value of the line or path integral is entirely dependent on the particular path taken from a to b; however, there are special class of vector functions for which the value of line integral entirely depends upon the end points. A vector function that possesses this property is dubbed as a conservative; otherwise, it is non-conservative. For instance, electrostatic force and gravitational force are examples of conservative forces.

Example 1.20 Evaluate the line integral for the function $\vec{V} = y^2\hat{x} + 2x(y + 1)\hat{y}$ along paths (1) and (2) as shown in Fig. 1.19

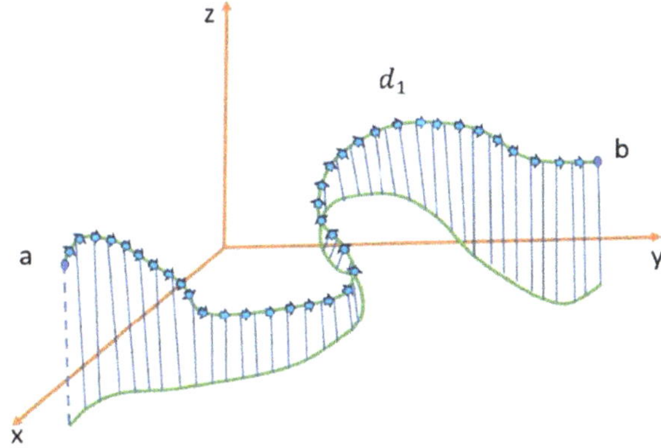

Fig. 1.18 Representation of a surface integral

Fig. 1.19 Path representation for line integral evaluation

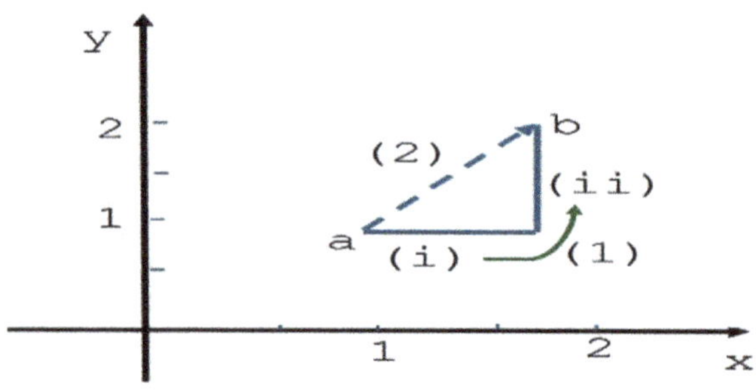

$$\vec{dl} = dx\hat{x} + dy\hat{y} + dz\hat{z}$$

Solution:

The path (1) has two parts (i) and (ii).
   Consider path (1)

(i) $\vec{dl} = dx\hat{x}$, $dy = 0$ and $y = 1$

$$\vec{V} \cdot \vec{dl} = y^2 dx = dx$$

$$\int \vec{V} \cdot \vec{dl} = \int_1^2 dx = 1$$

(ii) $\vec{dl} = dy\hat{y}$, $dx = 0$, $x = 2$ along y

$$\vec{V} \cdot \vec{dl} = 2x(y + 1)dy = 4(y + 1)dy$$

$$\int \vec{V} \cdot \vec{dl} = \int_1^2 4y\,dy + 4\int_1^2 dy = 10$$

Therefore, along path (1)

$$\int \vec{V} \cdot \vec{dl}_{(i)} + \int \vec{V} \cdot \vec{dl}_{(ii)} = 11$$

Along path (2)

$$\vec{dl} = dx\hat{x} + dy\hat{y} + dz\hat{z}$$
$$\vec{V} = y^2\hat{x} + 2x(y + 1)\hat{y}$$
$$x = y$$

Therefore $dx = dy$

$$\vec{V} \cdot \vec{dl} = x^2 dx + 2x(x+1)dx = 3x^2 dx + 2x dx$$

$$\int \vec{V} \cdot \vec{dl} = \int_1^2 3x^2 dx + \int_1^2 2x dx = 10$$

$$\oint \vec{V} \cdot \vec{dl} = 11 - 10 = 1$$

However, the value of integral depends upon the path; therefore, it is a non-conservative vector function.

1.25.2 *Surface Integral*

An expression of the following form is dubbed as a surface integral

$$\oint_S \vec{V} \cdot \vec{da}$$

where $\vec{V}$ is some vector function and $\vec{da}$ being some infinitesimal area element, with the direction being normal to the surface. For closed surface resembling with balloon, a circle is put on the integral sign as $\oint \vec{V} \cdot \vec{da}$. If the vector function under question represents a fluid flow, then the surface integral $\oint \vec{V} \cdot \vec{da}$ corresponds to the flux through the surface. Generally, the value of surface integral depends upon the particular surface chosen but we often encounter a special class of vector functions for which it is independent of the surface chosen and is entirely determined by the boundary line. Figure 1.20 illustrates the concept of the surface integral by depicting a vector field interacting with a surface. The surface is shown with differential area elements $\vec{da}$, which are oriented normal to the surface at each point. If the surface is closed, such as a sphere or an arbitrary enclosed shape, the integral accounts for the net flow of the vector field $\vec{V}$ across the surface.

Example 1.21 Evaluate the surface integral of the vector $\vec{V} = 2xz\hat{x} + (x+2)\hat{y} + y(z^3 - 3)\hat{z}$ on the following cube excluding its bottom (Fig. 1.21).

Solution:

Face (i): $x = 2$; $\vec{da} = dydz\hat{x}$

$$\vec{V} \cdot \vec{da} = 2xz dydz = 4z dydz$$

$$\int \vec{V} \cdot \vec{da} = 4 \int_0^2 z dz \int_0^2 dy = 16$$

Fig. 1.20 Surface element representation in 3-D space

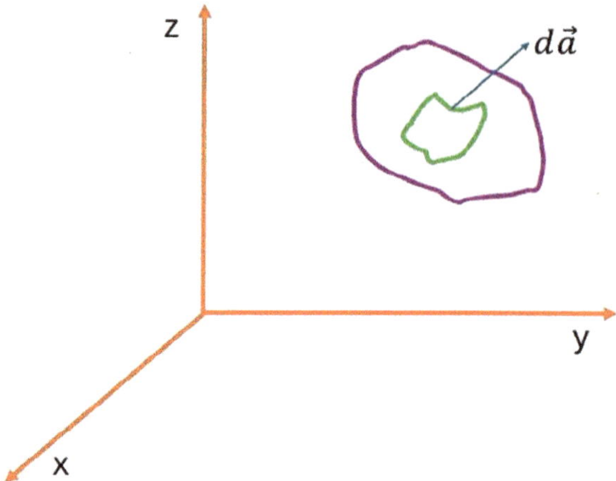

Fig. 1.21 Flux through a cubical surface in 3-D space

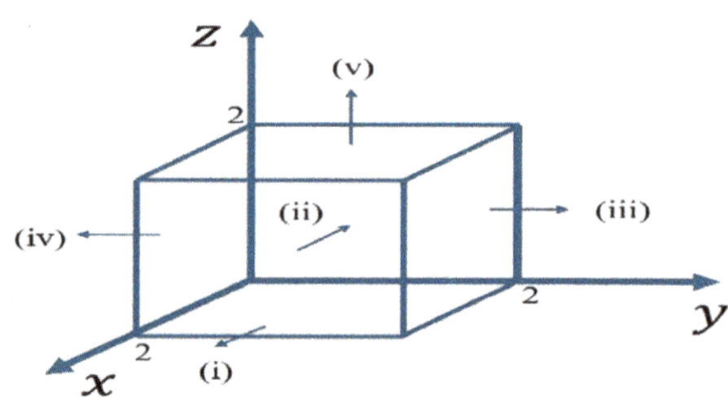

Face (ii):

$$x = 0, \mathrm{d}\vec{a} = \mathrm{d}y\mathrm{d}z(-\hat{x}) = -\mathrm{d}y\mathrm{d}z\hat{x}$$
$$\vec{V} \cdot \mathrm{d}\vec{a} = 2xz(-\mathrm{d}y\mathrm{d}z) = 0$$
$$\int \vec{V} \cdot \mathrm{d}\vec{a} = 0$$

Face (iii):

$$y = 2, \mathrm{d}\vec{a} = \mathrm{d}x\mathrm{d}z\hat{y}$$
$$\vec{V} \cdot \mathrm{d}\vec{a} = (x + 2)\mathrm{d}x\mathrm{d}z$$
$$\int \vec{V} \cdot \mathrm{d}\vec{a} = \int_0^2 (x + 2)\mathrm{d}x \int_0^2 \mathrm{d}z = 12$$

Face (iv):

$$y = 0;\ \mathrm{d}\vec{a} = -\mathrm{d}x\mathrm{d}z\hat{y}$$

$$\vec{V} \cdot \mathrm{d}\vec{a} = -(x + 2)\mathrm{d}x\mathrm{d}z$$

$$\int \vec{V} \cdot \mathrm{d}\vec{a} = -12$$

Face (v): $z = 2,\ d\vec{a} = dxdy\hat{z}$

$$\vec{V} \cdot \mathrm{d}\vec{a} = y\left(z^3 - 3\right)\mathrm{d}x\mathrm{d}y$$

$$\int \vec{V} \cdot \mathrm{d}\vec{a} = 4$$

The surface integral on all the surfaces

$$\oint_S \vec{V} \cdot \mathrm{d}\vec{a} = \oint_i \vec{V} \cdot \mathrm{d}\vec{a} + \oint_{ii} \vec{V} \cdot \mathrm{d}\vec{a} + \oint_{iii} \vec{V} \cdot \mathrm{d}\vec{a} + \oint_{iv} \vec{V} \cdot \mathrm{d}\vec{a} + \oint_{v} \vec{V} \cdot \mathrm{d}\vec{a}$$

$$\oint_S \vec{V} \cdot \mathrm{d}\vec{a} = 16 + 0 + 12 - 12 + 4 = 20$$

1.25.3 Volume Integral

If T is a scalar function, then volume integral of T is given by

$$\oint_V T\mathrm{d}\tau$$

where $\mathrm{d}\tau$ is an infinitesimal volume element. The volume element in terms of Cartesian coordinates can be written as under

$$\mathrm{d}\tau = \mathrm{d}x\mathrm{d}y\mathrm{d}z$$

It is noteworthy that if T is the density of a substance, then the volume integral would result in the mass of the substance. Moreover, we usually come across volume integrals of vector functions in physics.

$$\int \vec{V}\mathrm{d}\tau_V = \int \left(V_x\hat{x} + V_y\hat{y} + V_z\hat{z}\right)\mathrm{d}\tau = \hat{x}\int V_x\mathrm{d}\tau + \hat{y}\int V_y\mathrm{d}\tau + \hat{z}\int V_z\mathrm{d}\tau$$

Each component of the vector function is integrated separately over the volume, and the result is a vector quantity.

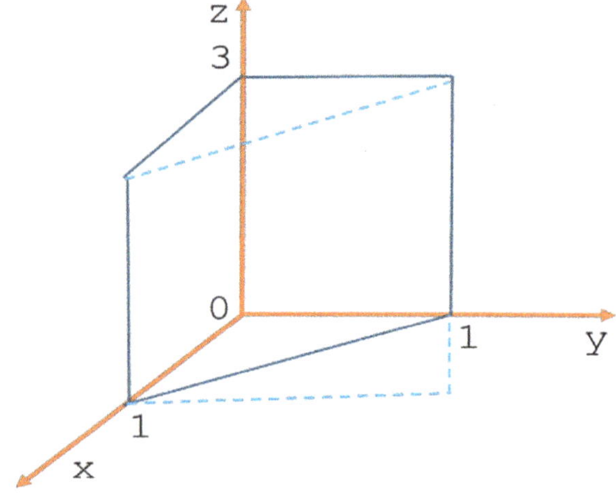

Fig. 1.22 A Prism represented in a three-dimensional Cartesian coordinate system

Example 1.22 Calculate the volume integral of the function $T = xyz^2$ over the prism as shown in Fig. 1.22

$$\int T \mathrm{d}\tau = \int xyz^2 \mathrm{d}\tau$$

Solution: Here we have to couple x and y

$$x : 0 \rightarrow 1 - y$$
$$y : 0 \rightarrow 1$$

The above integral can be written as:

$$\int T \mathrm{d}\tau = \int_0^3 z^2 \left\{ \int_0^1 y \left[\int_0^{1-y} x \mathrm{d}x \right] \mathrm{d}y \right\} \mathrm{d}z$$

$$\int T \mathrm{d}\tau = \int_0^3 z^2 \left\{ \int_0^1 \frac{yx^2}{2} \Big|_0^{1-y} \mathrm{d}y \right\} \mathrm{d}z$$

$$\int T \mathrm{d}\tau = \frac{1}{2} \int_0^3 z^2 \mathrm{d}z \int_0^1 (1-y)^2 y \mathrm{d}y$$

$$\int T \mathrm{d}\tau = \frac{3}{8}$$

1.26 Fundamental Theorem for Gradients

The fundamental theorem for gradients refers to a basic result in vector calculus that connects the gradient of a scalar field to the change in that scalar field along a curve. It is often referred as the gradient theorem or fundamental theorem. Consider some scalar function T. If we have indefinite integral, we can write:

$$\int dT = T(x) + c$$

$$\int_{a}^{b} dT = T(b) - T(a)$$

Any integrable function has its antiderivative (Fig. 1.23).

$$\int_{a}^{b} F(x) = F(b) - F(a) \tag{1.71}$$

$$dT = \vec{\nabla}T \cdot \vec{dl}$$

$$\vec{\nabla} = \left(\frac{\partial}{\partial x}\right)\hat{x} + \left(\frac{\partial}{\partial y}\right)\hat{y} + \left(\frac{\partial}{\partial z}\right)\hat{z}$$

$$\vec{dl} = dx\hat{x} + dy\hat{y} + dz\hat{z}$$

$$\vec{\nabla}T \cdot \vec{dl} = \left(\frac{\partial T}{\partial x}\right)dx + \left(\frac{\partial T}{\partial y}\right)dy + \left(\frac{\partial T}{\partial z}\right)dz = dT$$

$$\int_{a}^{b} dT = \int_{a}^{b} \vec{\nabla}T \cdot \vec{dl} = T(b) - T(a) \tag{1.72}$$

This is the fundamental theorem on gradients. Here we have converted the path dependence into path independence.

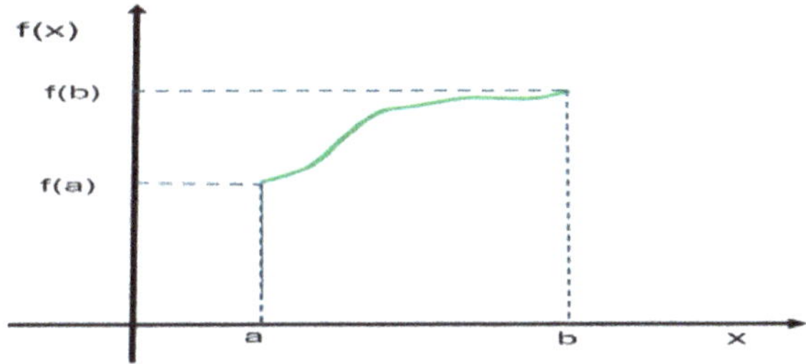

Fig. 1.23 Graphical representation of the fundamental theorem of calculus

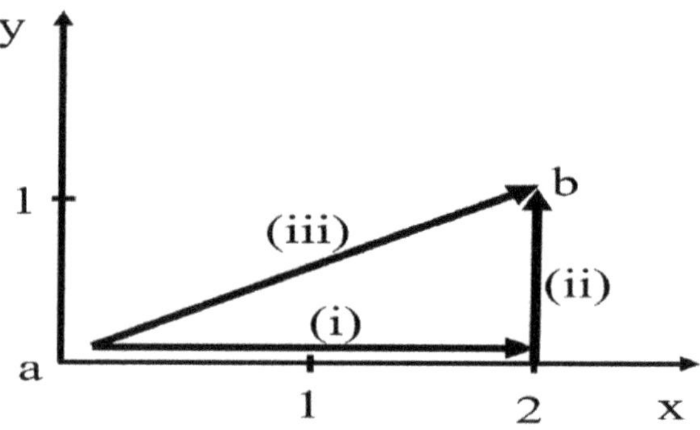

Fig. 1.24 Representation of a closed path integral in a two-dimensional coordinate system

Example 1.23 Evaluate the definite integral Here $F(x) = \frac{1}{x}$, if $F(x)$ is derivative of another function $\ln(x)$, then we can write

$$\int_1^3 \frac{1}{x}dx = \int_1^3 \ln(x)dx = \ln(3) - \ln(1) = \ln(3) = 1.1$$

Example 1.24 Let us assume that $T = xy^2$ is some function. Let the coordinates of points a and b are $(0, 0, 0)$ and $(2, 1, 0)$, respectively. Verify the fundamental theorem of gradients for this function (Fig. 1.24).

Here $T = xy^2$; a $(0, 0, 0)$ and b $(2, 1, 0)$

$$\vec{dl} = dx\hat{x} + dy\hat{y} + dz\hat{z}$$

$$\vec{\nabla}T = \left(\frac{\partial T}{\partial x}\right)\hat{x} + \left(\frac{\partial T}{\partial y}\right)\hat{y}$$

$$\vec{\nabla}T \cdot \vec{dl} = y^2 dx + 2xy dy$$

Path (i) $y = 0, \vec{dl} = dx\hat{x}$

$$\oint \vec{\nabla}T \cdot \vec{dl} = 0$$

Path (ii) $x = 2, dx = 0, \vec{dl} = dy\hat{y}$

$$\int \vec{\nabla}T \cdot \vec{dl} = \int 4ydy = \int_0^1 4ydy = 2$$

Therefore, sum along paths (i) and (ii) is given by

$$\int_a^b \vec{\nabla} T \cdot d\vec{l} = T(b) - T(a)$$

$$xy^2_{2,1|} - xy^2_{0,0|} = 2 - 0 = 2$$

Path (iii)

$$y = \frac{1}{2}x$$

$$dy = \frac{1}{2}dx$$

$$\int_a^b \vec{\nabla} T \cdot d\vec{l} = \int_0^2 \left(\frac{1}{4}x^2 + 2x\frac{1}{2}x\frac{1}{2}\right)dx$$

$$= \frac{3}{4}\int_0^2 x^2 dx = 2$$

This verifies the path independence of the above-mentioned function.

1.27   The Fundamental Theorem for Divergence

It relates volume integral to the surface integral. It has great importance in the classical electrodynamics. It states that the volume integral of divergence of some vector field, $\vec{V}$ over a certain volume is equal to the surface integral of the vector field, $\vec{V}$ over the closed surface bounding that volume.

Thus, integrating both sides over a curve from a to b

$$\oint_\tau \vec{\nabla} \cdot \vec{V} d\tau = \int_s \vec{V} \cdot d\vec{a} \tag{1.73}$$

The quantity $\vec{\nabla}.\vec{V}$ represents how much a physical quantity is diverging in a volume and therefore corresponds to the flux of that physical quantity escaping through the closed surface. Consider a spherical surface, with charge at the centre. There are two ways to make the calculation as per the above-mentioned theorem, either we can calculate volume integral or we can integrate surface integral. In both the ways we get the same answer.

Example 1.25   Verify divergence theorem for the function, $\vec{V} = y^2\hat{x} + (2xy + z^2)\hat{y} + 2yz\hat{z}$ and the unit cube situated at the origin.

   Solution:

$$\vec{V} = y^2\hat{x} + \left(2xy + z^2\right)\hat{y} + 2yz\hat{z}$$

$$\vec{\nabla}.\vec{V} = \left\{\left(\frac{\partial}{\partial x}\right)\hat{x} + \left(\frac{\partial}{\partial y}\right)\hat{y} + \left(\frac{\partial}{\partial z}\right)\hat{z}\right\}.\left\{y^2\hat{x} + \left(2xy + z^2\right)\hat{y} + 2yz\hat{z}\right\}$$

$$\vec{\nabla}.\vec{V} = 0 + 2x + 2y = 2x + 2y = 2(x + y)$$

$$\oint_V \vec{\nabla} \cdot \vec{V} d\tau = \oint_V 2(x + y)dxdydz = 2\int_0^1 \int_0^1 \int_0^1 (x + y)dxdydz$$

Since $x + y$ does not depend on Z, we first integrate over z

$$\int_0^1 dz = 1$$

Thus, the integral reduces to

$$2\int_0^1 \int_0^1 (x + y)dydx$$

Computing the inner integral

$$2\int_0^1 (x + y)dy = 2x + 1$$

Now, computing over x

$$\int_0^1 (2x + 1)dx = 2$$

Thus, the volume integral is

$$\oint_V \vec{\nabla} \cdot \vec{V} d\tau = 2$$

Now, the surface integral is

$$\int_s \vec{V} \cdot \hat{n} ds$$

The cube has six faces:

1. $x = 0$ (*Left*)
2. $x = 1$ (*Right*)
3. $y = 0$ (*Front*)
4. $y = 1$ (*Back*)
5. $z = 0$ (*Bottom*)
6. $z = 1$ (*Top*)

We will compute the flux through each face.

Face $x = 0$ (*normal $\hat{n} = -\hat{\imath}$*)

$$\vec{V} \cdot -\hat{\imath} = -y^2$$

$$\int_0^1 \int_0^1 -y^2 dy dz = -\frac{1}{3}$$

Face $x = 1$ (*normal $\hat{n} = \hat{\imath}$*)

$$\vec{V} \cdot \hat{\imath} = y^2$$

$$\int_0^1 \int_0^1 y^2 dy dz = \frac{1}{3}$$

Face $y = 0$ (*normal $\hat{n} = -\hat{\jmath}$*)

$$\vec{V} \cdot -\hat{\jmath} = -(2xy + z^2)$$

$$\int_0^1 \int_0^1 -z^2 dx dz = -\frac{1}{3}$$

Face $y = 1$ (*normal $\hat{n} = \hat{\jmath}$*)

$$\vec{V} \cdot \hat{\jmath} = (2xy + z^2)$$

$$\int_0^1 \int_0^1 (2x + z^2) dx dz = \frac{4}{3}$$

Face $z = 0$ (*normal $\hat{n} = -\hat{k}$*)

$$\vec{V} \cdot -\hat{k} = -2yz$$

$$\int_0^1 \int_0^1 -2yz\,\mathrm{d}x\mathrm{d}y = 0$$

Face $z = 1$ (*normal* $\hat{n} = \hat{k}$)

$$\vec{V}\hat{k} = 2yz$$

$$\int_0^1 \int_0^1 2y\,\mathrm{d}x\mathrm{d}y = 1$$

Hence, the total surface integral is

$$\frac{1}{3} - \frac{1}{3} - \frac{1}{3} + \frac{4}{3} + 0 + 1 = 2$$

Thus, the surface integral matches the volume integral, verifying the Divergence Theorem.

1.28   The Dirac Delta Function

The Dirac delta function, designated as, $\delta(x)$ is a fundamental concept in mathematics and physics, particularly in the fields of electrodynamics and quantum mechanics. Notwithstanding its name, it's not actually a "function" in the traditional sense, but rather a generalized function or distribution. We will provide its definition followed by a detailed discussion on its properties, applications and interpretation.

1.28.1   Concept and Intuitive Definition

The Dirac delta function is often described informally as a function that is zero everywhere except at $x = 0$, where it is infinitely high, in such a way that its total integral over the entire real line is 1, i.e.,

$$\int_{-\infty}^{\infty} \delta(x)\mathrm{d}x = 1 \tag{1.74}$$

The delta function can be perceived as the limit of a sequence of functions that peak more and more sharply at $x = 0$ while becoming narrower and maintaining a constant area under the curve equal to 1.

1.28.2 Mathematical Definition

Formally, the Dirac delta is defined by its action on a test function $f(x)$ under an integral:

$$\int_{-\infty}^{\infty} \delta(x) f(x) dx = f(0) \tag{1.75}$$

It is evident from the above integral that when Dirac delta is operated upon some test function $f(x)$, it results in the value of the function at $x = 0$. This is the key property of the delta function; essentially, it assumes the value of $f(x)$ at a specific point.

In more general terms, for any shift $a \in \mathbb{R}$

$$\int_{-\infty}^{\infty} \delta(x - a) f(x) dx = f(a) \tag{1.76}$$

The Dirac delta $\delta(x - a)$ is a delta function centred at $x = a$. It is zero everywhere except at $x = a$ and integrates to 1 over any interval that includes a.

1.28.3 Properties of the Dirac Delta Function

- **Shifting Property**: The delta function acts as a "shifting" function. Given a function $f(x)$ and the delta function $\delta(x - a)$, the integral

$$\int_{-\infty}^{\infty} \delta(x - a) f(x) dx = f(a) \tag{1.77}$$

  assumes the value of the function at $x = a$. This property makes the delta function useful in representing point sources or impulses.
- **Evenness**: The Dirac delta is an even function, i.e.,

$$\delta(x) = \delta(-x) \tag{1.78}$$

- **Scaling Property**: The Dirac delta scales inversely with the scaling of the variable. If α is a non-zero constant, then

$$\delta(\alpha x) = \frac{1}{|\alpha|}\delta(x) \tag{1.79}$$

- **Derivative**: The Dirac delta function can be differentiated in the sense of distributions. For example, the derivative of $\delta(x)$ denoted as $\delta'(x)$ satisfies

$$\int_{-\infty}^{\infty} \delta'(x)f(x)\mathrm{d}x = -f'(0) \tag{1.80}$$

Example 1.26 Evaluate the integral $I = \int_{-\infty}^{\infty} \delta(x-a)f(x)\mathrm{d}x$.
Solution:
By the shifting property of the Dirac delta function

$$\int_{-\infty}^{\infty} \delta(x-a)f(x)\mathrm{d}x = f(a)$$

Thus, the solution of the integral $I = f(a)$.

Example 1.26 Evaluate the following integral using the properties of the Dirac delta function

$$\int_{-\infty}^{\infty} \delta(x-2)f(x)\mathrm{d}x$$

where $f(x)$ is a continuous function.
Solution:

$$\int_{-\infty}^{\infty} \delta(x-2)f(x)\mathrm{d}x = f(2)$$

Example 1.27 Evaluate the integral

$$\int_{-\infty}^{\infty} \delta(3x-6)g(x)\mathrm{d}x$$

Solution: First, use the scaling property of the Dirac delta function

$$\delta(ax - b) = \frac{1}{|a|}\delta\left(x - \frac{b}{a}\right)$$

For the given problem

$$\delta(3x - 6) = \frac{1}{|3|}\delta(x - 2)$$

Now,

$$\int_{-\infty}^{\infty} \frac{1}{|3|}\delta(x - 2)g(x)\mathrm{d}x = \frac{1}{3}g(2)$$

1.29 Special Functions

In electrodynamics, the solutions to boundary value problems heavily depend on the symmetry of the system. For problems with spherical symmetry, such as potentials around spheres or charge distributions with angular dependence, Legendre functions naturally arise. These functions efficiently describe the angular behaviour of fields and potentials, making them indispensable in problems like the electric potential of a conducting sphere or the multipole expansion of charge distributions. For systems with cylindrical symmetry, such as infinitely long wires, coaxial cables, or waveguides, Bessel functions become essential. These functions characterize the radial dependence of fields in cylindrical geometries, providing solutions for problems like the potential around a charged cylinder or electromagnetic wave propagation in cylindrical waveguides. Together, they form the mathematical backbone for solving electrostatic and magnetostatic problems in these fundamental geometries, ensuring that boundary conditions are met and physical insights are preserved.

1.29.1 Legendre Polynomials

Legendre polynomials, denoted as $P_n(x)$ where n is a non-negative integer, are a sequence of orthogonal polynomials that play a significant role in mathematical physics, especially in solving problems with spherical symmetry, electrodynamics and quantum mechanics. They satisfy the Legendre differential equation:

$$\left(1 - x^2\right)\frac{d^2 P_n(x)}{dx^2} - 2x\frac{dP_n(x)}{dx} + n(n + 1)P_n(x) = 0$$

1.29.2　Properties of Legendre Polynomials

1. **Orthogonality**: Legendre polynomials are orthogonal on the interval $[-1, 1]$ with respect to the weight function $w(x) = 1$:

$$\int_{-1}^{1} P_m(x)P_n(x)\mathrm{d}x = 0 \quad \text{for} \quad m \neq n \tag{1.81}$$

When $m = n$, this integral gives a constant proportional to $\frac{2}{2n+1}$.

2. **Normalization**: The polynomials can be normalized so that $P_n(1) = 1$ for all n.
3. **Recurrence Relation**: Legendre polynomials satisfy a three-term recurrence relation, which allows the computation of higher-degree polynomials from lower ones:

$$(n + 1)P_{n+1}(x) = (2n + 1)xP_n(x) - nP_{n-1}(x) \tag{1.82}$$

With $P_0(x) = 1$ and $P_1(x) = x$ as initial conditions.

4. **Explicit Form**: The Legendre polynomials can be represented explicitly using Rodrigue's formula:

$$P_n(x) = \frac{1}{2^n n!} \frac{\mathrm{d}^n}{\mathrm{d}x^n}\left(x^2 - 1\right)^n \tag{1.83}$$

This expression is useful for deriving specific polynomial terms and their properties.

5. **Orthogonality and Completeness in Function Expansion**: Legendre polynomials form a complete basis for representing functions defined on $[-1, 1]$. Any reasonable function $f(x)$ defined in this interval can be expanded as a series:

$$f(x) = \sum_{n=0}^{\infty} a_n P_n(x) \tag{1.84}$$

where the coefficients a_n are determined by projecting $f(x)$ onto each $P_n(x)$ while using the orthogonality property.

6. **Parity**: $P_n(x)$ is an even function if n is even and an odd function if n is odd:

$$P_n(-x) = (-1)^n P_n(x) \tag{1.85}$$

7. **Generating Function**: Legendre polynomials have a generating function, which is especially useful in deriving properties and summing series:

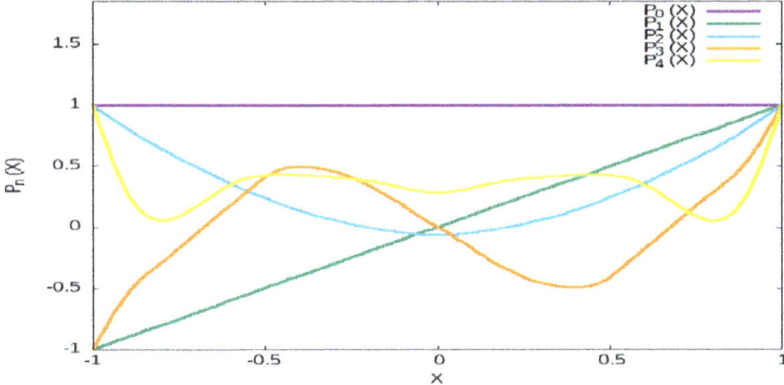

Fig. 1.25 Schematic representation of Legendre polynomials

$$\frac{1}{\sqrt{1 - 2xt + t^2}} = \sum_{n=0}^{\infty} P_n(x)t^n \tag{1.86}$$

Figure 1.25 shows the first five Legendre polynomials, $P_n(x)$ which are solutions to Legendre's differential equation. The x-axis represents the variable x ranging from -1 to 1, and the y-axis represents the value of the Legendre polynomials, $P_n(x)$. The Legendre polynomials are orthogonal over the interval $[-1, 1]$ and are frequently used in solving problems involving spherical symmetry in physics, such as electrostatic and gravitational potentials.

From Fig. 1.25 $P_0(x)$ (in purple) is the 0th Legendre polynomial, which is a constant function equal to 1. $P_1(x)$ (in green) is the 1st Legendre polynomial, which is a linear function given by $P_1(x) = 1$. $P_2(x)$ (in light blue) is the 2nd Legendre polynomial, a quadratic function $P_2(x) = \frac{1}{2}(3x^2 - 1)$. $P_3(x)$ (in brown) is the 3rd Legendre polynomial, a cubic function $P_3(x) = \frac{1}{2}(5x^3 - 3x^2)$. $P_4(x)$ (in yellow) is the 4th Legendre polynomial, a quartic function $P_4(x) = \frac{1}{8}(35x^4 - 30x^2 + 3)$.

1.29.3 Bessel Functions

Bessel functions are a family of solutions to Bessel's differential equation, which appears in a wide range of physics and engineering problems, particularly those involving cylindrical or spherical symmetry. Named after the mathematician Friedrich Bessel, these functions are indispensable in analysing wave propagation, potential problems and many other areas of applied mathematics.

The standard form of Bessel's differential equation is:

$$x^2 \frac{d^2 y}{dx^2} + x \frac{dy}{dx} + (x^2 - n^2)y = 0 \tag{1.87}$$

where n is a real or complex constant, often referred as the order of the Bessel function. Solutions to this equation, known as Bessel functions, are classified primarily into two main types: Bessel functions of the first kind $J_n(x)$ and Bessel functions of the second kind $Y_n(x)$.

1.29.4 Properties of Bessel Functions

1. **Orthogonality**: Bessel functions of different orders $J_n(x)$ are orthogonal with respect to a weighted inner product:

$$\int_0^1 x J_n(a_{m,n}, x) J_n(a_{k,n}, x)\mathrm{d}x = 0 \quad \text{for} \quad m \neq k \tag{1.88}$$

where $a_{m,n}$ are the roots of J_n. Orthogonality is useful in solving boundary value problems in cylindrical coordinates.

2. **Recurrence Relations**: Bessel functions satisfy several recurrence relations. For example,

$$J_{n+1}(x) + J_{n-1}(x) = \frac{2n}{x} J_n(x)$$

$$J_{n-1}(x) - J_{n+1}(x) = 2J'(x) \tag{1.89}$$

These relations allow computations of higher-order Bessel functions from lower-order ones.

3. **Zeros of Bessel Functions**: The zeros of $J_n(x)$ are of interest in many physical applications, such as the vibration modes of a circular membrane. For each order n, the function $J_n(x)$ has an infinite number of positive real zeros, which are typically denoted by $J_{n+k}(x)$ for the kth zero of $J_n(x)$.

4. **Integral Representations or Integer Order n:** Bessel functions can also be represented by integrals. For instance, the Bessel function $J_n(x)$ has an integral representation:

$$J_n(x) = \frac{1}{\pi} \int_0^\pi \cos(nt - x \sin t)\mathrm{d}t$$

Integral representations are useful for theoretical derivations and numerical computations.

Figure 1.26 is a plot of Bessel functions of the first kind, denoted as $J_n(x)$, for different integer orders n ranging from 0 to 4.

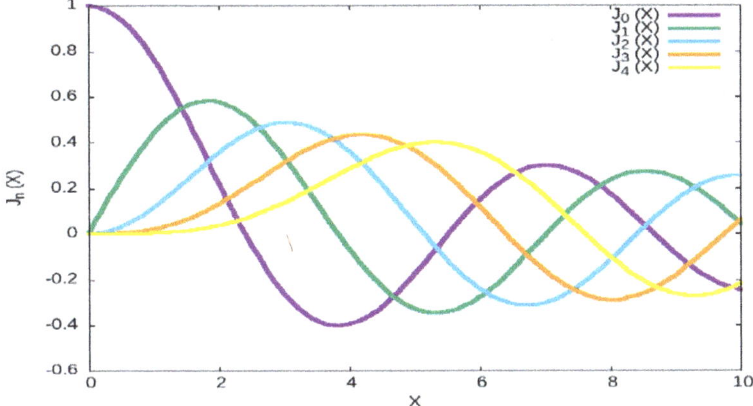

Fig. 1.26 Schematic representation of Bessel functions of first kind

Unsolved Problems:

Problem 1.1 Four points A (1, 2, 3), B (2, 0, 1), C (0, − 1, 4) and D (1, 1, 0) form a tetrahedron. Find the volume of this tetrahedron using vectors.
  Ans. $\frac{7}{3}$unit^3.

Problem 1.2 Consider the vector field $\vec{F}(x, y, z) = \left(y^2 z\hat{i} + x^2 z\hat{j} + xy^2\hat{k}\right)$.

 (i) Compute the curl of the vector field $\vec{F}(x, y, z)$. Ans. $\left(2xy - x^2\right)\hat{i} + (2xz - 2yz)\hat{k}$.
(ii) Evaluate the curl at the point (1, 2, 3).

   Ans. $3\hat{i} - 6\hat{k}$

Problem 1.3 Verify Gauss divergence theorem for the vector field $\vec{F} = x^2\hat{i} + y^2\hat{j} + z^2\hat{k}$ over the surface of a cube with side of length 2, centred at origin (i.e., $x, y, z \epsilon[-1, 1]$).

Problem 1.4 Evaluate the line integral for the scalar field $f(x, y) = x^2 + y^2$ along the straight line from (0, 0) to (1, 1).
  Ans. $\frac{2\sqrt{2}}{3}$

Problem 1.5 Evaluate the integral
$$I = \int\limits_{-\infty}^{\infty} x^2 \delta\left(x^2 - 1\right) \mathrm{d}x$$
Ans. 1

Problem 1.6 Given that force $\vec{F} = 4\hat{i} + 5\hat{j}N$ and displacement $\vec{d} = 5\hat{i} + 12\hat{j}m$. Calculate the work done W by a force $\vec{F}$ over a displacement $\vec{d}$.
  Ans. 80 J.

Problem 1.7 The vectors $\vec{A} = 3\hat{i} + 4\hat{j}$ and $\vec{B} = -5\hat{i} + 2\hat{j}$ are added to form a resultant vector $\vec{C} = \vec{A} + \vec{B}$. Find the magnitude of $\vec{C}$ using law of cosines.
  Ans. 6.32.

Problem 1.8 An electric dipole with dipole moment $\vec{p} = 4\hat{i}\,Cm$ is placed in a uniform electric field $\vec{E} = 10\hat{j}\,N/C$. Calculate the torque acting on the dipole.

Ans. $40\hat{k}$ Nm.

Problem 1.9 The vectors $\vec{A} = 3\hat{i} + 4\hat{j} + \hat{k}$ and $\vec{B} = \hat{i} - 2\hat{j} + 2\hat{k}$ form a parallelogram. Find the area of the parallelogram.

Ans. 15 units^2

Problem 1.10 Calculate the gradient of $\phi(\vec{r}) = \frac{1}{|\vec{r}|}$ and show that it aligns with the field of a point charge, i.e., $\vec{\nabla}\left(\frac{1}{|\vec{r}|}\right) = -\frac{\vec{r}}{|\vec{r}|^3}$.

Problem 1.11 Evaluate the integral $\int_{-\infty}^{\infty} (3x + 2)\delta(x - 1)\mathrm{d}x$

Ans. 5

Problem 1.12 Use the divergence theorem to evaluate the flux of the vector $\vec{F} = \frac{\hat{r}}{r^2}$ over the surface of a sphere of radius R centred at the origin.

Ans. 4π.

Problem 1.13 Calculate the gradient of the scalar field $(\vec{r}) = \frac{1}{4\pi\,\epsilon_0 r}$ and interpret it in terms of the Dirac delta function.

Ans. $\delta(\vec{r})$.

Problem 1.14 Consider the vector field $\vec{G} = z\hat{i} - x\hat{i}$. Show that $\vec{\nabla}.\vec{G} = 0$ and verify that $\oint_S \vec{G} \cdot \mathrm{d}\vec{A} = 0$ for any closed surface S.

Problem 1.15 Show that $\int_{-1}^{+1} P_n(x)\mathrm{d}x = 0, \quad n \neq 0$.

and $\int_{-1}^{+1} P_n(x)\mathrm{d}x = 2, \quad n = 0$.

Problem 1.16 Let $P_n(x)$ be the Legendre polynomial of degree n. Show that for any function $f(x)$, for which the nth derivative is continuous.

$$\int_{-1}^{+1} f(x)P_n(x)\mathrm{d}x = \frac{(-1)^n}{2^n n!} \int_{-1}^{+1} (x^2 - 1)^n f^n(x)\mathrm{d}x$$

Problem 1.17 Show that $J_{1/2}(x) = \sqrt{\left(\frac{2}{\pi x}\right)} \sin(x)$ and $J_{-1/2}(x) = \sqrt{\left(\frac{2}{\pi x}\right)} \cos(x)$.

Problem 1.18 Prove that $\sqrt{\left(\frac{\pi x}{2}\right)} J_{3/2}(x) = \frac{\sin(x)}{x} - \cos(x)$.

1.30 Summary

- **Definition of a Vector**: A vector is defined as a mathematical object with both magnitude and direction, differentiating it from scalars, which possess only magnitude. Vectors are vital for describing physical quantities requiring both properties, such as force and velocity.

- **Vector Notation and Components**: Vectors are represented symbolically by $\vec{A}$ or by bold-faced letter **A.** On splitting it into components along the coordinate axes, each component reflects the vector's magnitude along each direction.
- **Negative of a Vector**: The negative vector $-\vec{A}$ has the same magnitude as $\vec{A}$ but points in the opposite direction. This concept is fundamental in vector algebra, especially for operations like subtraction.
- **Graphical Interpretation of Vectors**: Vectors can be visualized as arrows, where the length represents magnitude, and direction is indicated by the arrowhead. Graphical representation aids in understanding vector operations such as addition.
- **Vector Addition**: Adding vectors involves placing them head-to-tail to determine a resultant vector. This operation, crucial in physics and engineering, represents the combined effect of two or more vectors.
- **Scalar and Vector Multiplication**

  - **Scalar Multiplication**: Involves multiplying a vector by a scalar, scaling its magnitude without changing its direction.
  - **Dot Product (Scalar Product)**: Provides a scalar result that represents the projection of one vector onto another, used in work and energy calculations.
  - **Cross Product (Vector Product)**: Results in a vector perpendicular to the plane of the original vectors, essential in calculating torque and angular momentum.

- **Triple Products**

  - **Scalar Triple Product**: Determines the volume of a parallelepiped formed by three vectors.
  - **Vector Triple Product**: Represents a vector perpendicular to the plane defined by two vectors, simplifying complex vector operations.

- **Transformation of Vectors**: Vector components adjust with changes in coordinate systems through transformation matrices. Tensors extend this concept, remaining invariant under such transformations.
- **Kronecker Delta and Levi–Civita Symbol**

  - **Kronecker Delta**: Serves as a unit matrix in tensor notation, simplifying vector operations.
  - **Levi–Civita Symbol**: Facilitates defining cross products and tensor properties, essential for advanced vector manipulation.

- **Differential Calculus of Vectors**

  - **Gradient, Divergence and Curl**: Key operations that describe how vectors vary in space, essential for analysing fields in physics.
  - **Line, Surface and Volume Integrals**: Represent physical phenomena like work, flux and charge distributions in space.

- **Dirac Delta Function**: Acts as a generalized function to represent point sources or impulses in physics, central to electrodynamics and quantum mechanics.

- **Special Functions**

  - **Legendre Polynomials**: Employed in spherical symmetry problems, aiding in solutions to differential equations.
  - **Bessel Functions**: Arise in cylindrical symmetry problems, relevant in wave propagation and electromagnetism.

- This chapter provides a comprehensive introduction to vectors, vector operations and specialized functions, laying the groundwork for applications in electrodynamics involving spatial and directional quantities.

Chapter 2
Boundary Value Problems-I

Abstract This chapter focuses on the Laplace equation, $\nabla^2 V = 0$, a fundamental equation in electrostatics and electrodynamics. Solutions in spherical coordinates are explored using the separation of variable technique, dividing the problem into radial, polar and azimuthal components. Key concepts include Legendre polynomials, derived from the Legendre differential equation and Rodrigue's formula, which naturally emerge in systems with spherical symmetry. Boundary conditions are analysed in both spherical and cylindrical coordinate systems. In spherical symmetry, solutions address potentials inside or outside spherical conductors, while cylindrical symmetry involves cylindrical geometries and often employs Bessel functions. Practical applications demonstrate the Laplace equation's utility in determining electrostatic potentials under specified boundary values or symmetries, such as potentials on spherical shells or cylindrical conductors. The chapter also covers generating functions and recursion relations for Legendre polynomials, which are critical in solving boundary value problems with spherical symmetry. The First and Second Uniqueness Theorems are demonstrated, ensuring the uniqueness of solutions to the Laplace equation under given boundary or charge distribution conditions, reinforcing its strength in physical applications.

Keywords Laplace equation · Spherical and cylindrical symmetry · Bessel functions and uniqueness theorems

2.1 Introduction

A special class of problems that we often encounter in electrostatics involve particular boundary conditions. Pertinently, depending on the scenario, either the electric potential or the surface charge density may be relevant for consideration, is fixed on the boundary surface involved in these problems. These problems are often solved explicitly by employing the technique of Green's functions. However, the process is not too handy always. From time to time, a number of techniques have been employed to evaluate boundary value problems. The various techniques that have been formulated to solve boundary value problems are briefly summarized as follows.

2.1.1 *The Approach of Using Images*

This method is particularly useful for problems involving one or more-point charges in the vicinity of boundary surfaces. Notably, under stable conditions, it is often possible to deduce from the geometry of the system that a certain number of charges, positioned outside the region of interest, can be selected with appropriate magnitudes to satisfy the boundary conditions. These charges are referred as image charges and are, therefore, relevant in the usage of various image-based problems. This method or technique is depicted as the method of images.

2.1.2 *The Method of Orthogonal Expansion*

This method involves the solution of differential equations expressed as an expansion in orthogonal functions. This is a powerful technique employed in a wide variety of problems in electrostatics. The orthogonal set chosen depends on symmetries involved in a particular problem.

2.1.3 *The Finite Element Analysis (FEA) Technique*

This technique involves a wide variety of numerical methods for addressing boundary value problems that we usually encounter in physics and engineering.

2.2 The Laplace Equation

The Laplace equation is a cornerstone in physics, serving as one of the most fundamental and pivotal mathematical expression in widespread physical processes which is the solution to a wide variety of problems occurring in electrodynamics and thermodynamics. Therefore, it is worthwhile to know how to solve this equation in various coordinate systems. We will try to solve this equation in spherical polar coordinate system. Pertinently, for the region or space where the charge density is zero, Poisson equation reduces to the Laplace equation.

$$\nabla^2 V = \frac{1}{r^2}\frac{\partial}{\partial r}\left(r\frac{\partial V}{\partial r}\right); \; V(r, \theta, \phi) \tag{2.1}$$

Fig. 2.1 Schematic representation of spherical polar coordinates

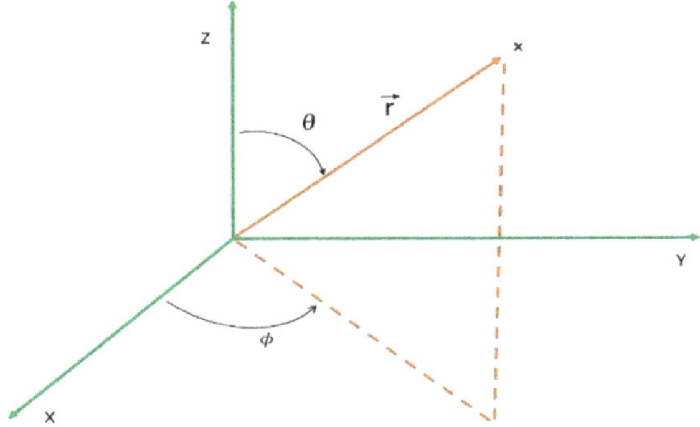

Re-writing the above equation in spherical polar coordinates, we get

$$\nabla^2 V = \frac{1}{r^2}\frac{\partial}{\partial r}\left(r^2\frac{\partial V}{\partial r}\right) + \frac{1}{r^2\sin(\theta)}\frac{\partial}{\partial\theta}\left(\sin(\theta)\frac{\partial V}{\partial\theta}\right)$$

$$+ \frac{1}{r^2\sin^2(\theta)}\frac{\partial^2 V}{\partial\phi^2} = 0 \tag{2.2}$$

where θ being the polar angle and ϕ the azimuthal angle. Using variable separation technique (Fig. 2.1).

$$V(r,\theta,\phi) = \frac{U(r)}{r}P(\theta)Q(\phi) \tag{2.3}$$

We take $\frac{U(r)}{r}$ to make solution of Laplace equation easy. $U(r)$, $P(\theta)$ and $Q(\phi)$ are potentials, which depends on r, θ and ϕ, respectively.

$$rV(r,\theta,\phi) = U(r)P(\theta)Q(\phi)$$

Multiplying Eq. (2.2) with $\frac{r^3\sin^2(\theta)}{UPQ}$ throughout, we get

$$r^2\sin^2(\theta)\frac{1}{U}\frac{d^2U}{dr^2} + \frac{\sin(\theta)}{P}\frac{d}{d\theta}\left(\sin(\theta)\frac{dP}{d\theta}\right) + \frac{1}{Q}\frac{d^2Q}{d\phi^2} = 0 \tag{2.4}$$

The first term in Eq. (2.4) depends on both r and θ, the second term solely depends on θ, however, the third term exclusively depends on ϕ.

Suppose the first two terms of Eq. (2.4) are equal to m^2 and the last term is equal to $-m^2$

$$\frac{1}{Q}\frac{d^2Q}{d\phi^2} = -m^2$$

The resolution of this differential equation yields

$$Q_m(\phi) = e^{\pm im\phi} = \cos(m\phi) \pm i \, \sin(m\phi)$$

These are the periodic functions, which means potential on the surface of sphere must be same; i.e., m must be an integer to force potential (Q) to be single valued (Fig. 2.2).

$$\phi = \phi \pm 2m\pi$$

$$r^2 \sin^2(\theta) \frac{1}{U} \frac{d^2 U}{dr^2} + \frac{\sin(\theta)}{P} \frac{d}{d\theta}\left(\sin(\theta)\frac{dP}{d\theta}\right) = m^2 \qquad (2.5)$$

Dividing the above equation by $\sin^2(\theta)$, we get

$$\frac{r^2}{U} \frac{d^2 U}{dr^2} + \frac{1}{P \sin(\theta)} \frac{d}{d\theta}\left(\sin(\theta)\frac{dP}{d\theta}\right) - \frac{m^2}{\sin^2(\theta)} = 0 \qquad (2.6)$$

The first term of Eq. (2.6) is function of r, the second term depends on θ and the last term is exclusively dependent on (θ, ϕ).

Put first term of Eq. (2.6) equal to α^2 and sum of second and third terms equal to $-\alpha^2$

$$\frac{r^2}{U} \frac{d^2 U}{dr^2} = \alpha^2$$

$$\frac{d^2 U}{dr^2} = \frac{\alpha^2}{r^2} U \qquad (2.7)$$

Fig. 2.2 A spherical distribution of charge Q contained inside a grounded conducting sphere of radius a

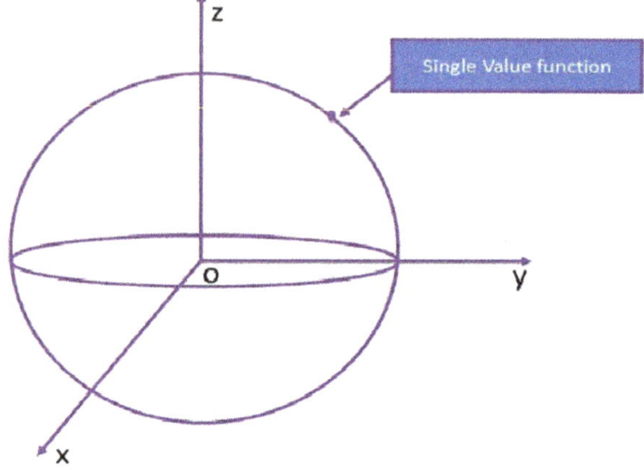

Consider the power series solution of Eq. (2.7) as follows:

$$U \sim r^{\rho} \tag{2.8}$$

Differentiate equation twice with respect to ρ, we get

$$U'' \sim \rho(\rho - 1)r^{\rho-2} \tag{2.9}$$

Substitute the expressions from Eqs. (2.8) and (2.9) into Eq. (2.7), we get

$$\rho(\rho - 1)r^{\rho-2} = \frac{\alpha^2}{r^2}r^{\rho} = \alpha^2 r^{\rho-2}$$

Let,

$$\rho(\rho - 1) = \alpha^2 \tag{2.10}$$

Both ρ and α are undetermined constants.

In order to simplify our calculation, suppose $\alpha^2 = l(l + 1)$, where l being the constant

$$\rho(\rho - 1) = \alpha^2 = l(l + 1)$$
$$\rho^2 - \rho - l(l + 1) = 0 \tag{2.11}$$

Which is the quadratic equation in ρ and hence solving it, we get

$$\rho^+ = l + 1; \; \rho^- = -l$$

As a result, the solution corresponding to the radial component is:

$$U(r) = Ar^{\rho^+} + Br^{\rho^-} \tag{2.12}$$

But we know $V(r) = \frac{U}{r}$.

Therefore,

$$V(r) = Ar^l + Br^{-(l+1)} \tag{2.13}$$

This is the solution of the radial part, where l is still to be determined.

Consider the second and third terms of Eq. (2.6)

$$l(l + 1) + \frac{1}{P\sin(\theta)}\frac{d}{d\theta}\left(\sin(\theta)\frac{dP}{d\theta}\right) - \frac{m^2}{\sin^2(\theta)} = 0 \tag{2.14}$$

Multiplying Eq. (2.14) by P throughout, we get

$$\frac{1}{\sin(\theta)}\frac{d}{d\theta}\left(\sin(\theta)\frac{dP}{d\theta}\right) + \left[l(l+1) - \frac{m^2}{\sin^2(\theta)}\right]P = 0 \qquad (2.15)$$

Which is the generalized Legendre differential equation. In order to simplify this equation, we consider a normalized unit circle. Let us consider the following variable transformation:

$$x = \cos(\theta); \, y = \sin(\theta)$$

By changing the rule of differentiation, we get

$$\frac{d}{d\theta} = \frac{dx}{d\theta}\frac{d}{dx} = -\sin(\theta)\frac{d}{dx} \qquad (2.16)$$

Multiplying Eq. (2.16) by $\sin(\theta)$ throughout, we get

$$\sin(\theta)\frac{d}{d\theta} = -\sin^2(\theta)\frac{d}{dx} = -\left(1 - \cos^2(\theta)\right)\frac{d}{dx}$$

$$\sin(\theta)\frac{d}{d\theta} = -\left(1 - x^2\right)\frac{d}{dx}; \quad \frac{1}{\sin(\theta)}\frac{d}{d\theta} = -\frac{d}{dx} \qquad (2.17)$$

Using Eq. (2.17) in Eq. (2.15), we get

$$\frac{d}{dx}\left[\left(1 - x^2\right)\frac{dP}{dx}\right] + \left[l(l+1) - \frac{m^2}{1 - x^2}\right]P = 0 \qquad (2.18)$$

This is also generalized Legendre differential equation, but in different style. This equation is very difficult to solve; therefore, we consider azimuthal symmetry.

2.2.1 Azimuthal Symmetry

To make calculations easy and simple, we assume azimuthal symmetry, meaning that our parameter V does not dependent on ϕ. Simply put, the partial derivative of V with respect to ϕ equal to zero. Considering the below-mentioned equation

$$\frac{d}{dx}\left[\left(1 - x^2\right)\frac{dP}{dx}\right] + l(l+1)P = 0 \qquad (2.19)$$

Re-writing the above equation, we obtain

$$\left(1 - x^2\right)\frac{d^2P}{dx^2} - 2x\frac{dP}{dx} + l(l+1)P = 0 \qquad (2.20)$$

When $x = \cos(\theta)$; $0 \leq \theta \leq \pi$ or $1 \leq \theta \leq -1$.
Further, Eq. (2.20) can be written as follows:

$$\frac{d^2P}{dx^2} - x^2\frac{d^2P}{dx^2} - 2x\frac{dP}{dx} + l(l+1)P = 0 \qquad (2.21)$$

This is the Legendre differential equation with azimuthal symmetry. Re-writing Eq. (2.21) while using P, P' and P'', it can exhibit as follows:

$$P'' - x^2P'' - 2xP' - 2xP' + l(l+1)P = 0 \qquad (2.22)$$

Assuming power series solution of Legendre differential equation

$$P(x) = a_0 + a_1x + a_2x^2 + a_3x^3 + a_4x^4 + \cdots + a_nx^n \qquad (2.23)$$

Differentiate above equation with respect to x, we get

$$P'(x) = a_1 + 2a_2x + 3a_3x^2 + 4a_4x^3 + \cdots + na_nx^{n-1} \qquad (2.24)$$

Further, differentiating Eq. (2.24) with respect to x, we get

$$P''(x) = 2a_2 + 6a_3x + 12a_4x^2 + \cdots + n(n-1)a_nx^{n-2} \qquad (2.25)$$

In order to solve Eq. (2.22), all the coefficients should be zero. To make calculation easy and simple, we will arrange the coefficients in the tabular form as under.

	Constant	x	x^2	x^3	$\ldots x^n$
P''	$2a_2$	$6a_3$	$12a_4$	$20a_5$	$(n+1)(n+2)a_{n+2}$
$-x^2P''$	–	–	$-2a_2$	$-6a_3$	$-n(n-1)a_n$
$-2xP'$	–	$-2a_1$	$-4a_2$	$-6a_3$	$-2na_n$
$l(l+1)P$	$l(l+1)a_0$	$l(l+1)a_1$	$l(l+1)a_2$	$l(l+1)a_3$	$l(l+1)a_n$

Thus, to solve Eq. (2.22), we insert Eqs. (2.23), (2.24), and (2.25) in Eq. (2.22). Further, we equate all the coefficients with zero. Let us equate constant term with zero, we get.

$$2a_2 + l(l+1)a_0 = 0$$

Solving it further, we get

$$a_2 = -\frac{l(l+1)a_0}{2} \qquad (2.26)$$

Further, we equate the coefficient of x with zero, and therefore, we obtain.

$$a_3 = \frac{-(l-1)(l+2)a_1}{3!} \tag{2.27}$$

Furthermore, we equate the coefficient of x^2 with zero and solving it, we get

$$a_4 = \frac{l(l+1)(l-2)(l+3)a_0}{4!} \tag{2.28}$$

Similarly, we equate the coefficient of x^3 with zero and solving it further, we get

$$a_5 = \frac{(l-1)(l+2)(l-3)(l+4)a_1}{5!} \tag{2.29}$$

Proceeding in the same manner we equate the coefficient of x^n with zero and, therefore, we get.

$$a_{n+2} = \frac{(n-l)(n+l+1)a_n}{(n+1)(n+2)} \tag{2.30}$$

Which is the recursion relation, it can produce all the coefficients.

If we start from $n = 0$, we will generate even series as $a_0, a_2, a_4,\ldots$ and if we start from $n = 1$, we will generate odd series as $a_1, a_3, a_5,\ldots$. Consequently, solutions to the second-order Legendre differential equation is the combination of even and odd series. We need to determine $a_0, a_1, \ldots$ from the initial conditions.

$$P(x) = a_0 + a_2x^2 + a_4x^4 + a_1x + a_3x^3 + +a_5x^5 + \cdots + a_nx^n \tag{2.31}$$

Further, we solve it and re-write it as follows:

$$P(x) = a_0\left[1 - \frac{l(l+1)x^2}{2} + \frac{l(l+1)(l-2)(l+3)x^4}{4!} + \cdots\right]$$
$$+ a_1\left[x - \frac{(l-1)(l+2)x^3}{3!} + \frac{(l-1)(l+2)(l-3)(l+4)x^5}{5!} + \cdots\right] \tag{2.32}$$

The solution converges for $x^2 < 1$ and diverges for $x^2 \geq 1$. Consider a_1 series

$$x - \frac{(l-1)(l+2)x^3}{3!} + \frac{(l-1)(l+2)(l-3)(l+4)x^5}{5!} + \cdots \tag{2.33}$$

For $l = 0$ and $x = 1$

$$x - \frac{(l-1)(l+2)x^3}{3!} + \frac{(l-1)(l+2)(l-3)(l+4)x^5}{5!} + \cdots = 1 + \frac{1}{3} + \frac{1}{5} + \cdots \tag{2.34}$$

Since a_1 series diverges and a_0 series converges. This can be shown using integral test.

$P(x) = a_0$, for $l = 0$, in case of even solutions

On the other hand, for $l = 1$ and $x = 1$. a_0 series diverges and a_1 stops at $a_1 x$.

$$P(x) = a_1 x, \text{ for } l = 1,$$

in case of odd solutions

We will not consider the negative values of l, because it will reproduce earlier results. For example, $l = -2$ will reproduce $P(x) = a_1 x$.

Therefore, the set of Legendre polynomial will be generated from Rodrigue's formula given by

$$P_l(x) = \frac{1}{2^l l!} \left(\frac{d}{dx}\right)^l (x^2 - 1)^l \tag{2.35}$$

However, for $x = 0$, 1 and 2, we get from above equation

$$P_0(x) = 1; P_1(x) = x, ; P_2(x) = \frac{1}{2}(3x^2 - 1)$$

Further, if we substitute $x = \cos(\theta)$ in above, we get

$$P_0(\cos(\theta)) = 1; P_1(\cos(\theta)) = \cos(\theta) \text{ and}$$
$$P_2(\cos(\theta)) = \frac{1}{2}(3\cos^2(\theta) - 1) \tag{2.36}$$

Therefore, we conclude that the solution of Legendre differential equation with azimuthal symmetry is given by:

$$V(r, \theta) = \sum_{l=0}^{\infty} \left(A_l r^l + B_l r^{-(l+1)}\right) P_l(\cos\theta) \tag{2.37}$$

In order to determine $V(r, \theta)$, we have to determine A_l, B_l and $P_l(\cos\theta)$.

Properties of Legendre Polynomials:

1. Legendre equation with azimuthal symmetry looks like an eigenvalue problem as follows:

$$\frac{d}{dx}\left[(1 - x^2)\frac{dP}{dx}\right] = -l(l + 1)P \tag{2.38}$$

However, the above equation can be rewritten as follows:

$$\hat{L}P = \Lambda P \tag{2.39}$$

where

$$\hat{L} = \frac{d}{dx}\left[(1-x^2)\frac{d}{dx}\right] \quad \text{and} \quad \Lambda = -l(l+1) \tag{2.40}$$

Furthermore, we can prove that $\hat{L} = \hat{L}^\dagger$; i.e., $\hat{L}$ is Hermitian operator and has therefore, real Eigen values given by $l(l+1)$.

2. The eigenfunctions associated with different eigenvalues are orthonormal.

$$\int_{-1}^{1} P_{l'}(x)P_l(x)dx = \frac{2}{2l+1}\delta_{l'l} \tag{2.41}$$

Example 2.1 Let a conducting spherical surface of radius 'a' with a potential $V(\theta)$ defined on its surface. Assume there are no free charges present within or outside the sphere. Using the appropriate boundary conditions deduce the electric potential everywhere within the sphere, particularly at the origin.

Solution:

The solution of the Legendre differential equation with azimuthal symmetry is exhibited as

$$V(r,\theta) = \sum_{l=0}^{\infty}\left(A_l r^l + B_l r^{-(l+1)}\right)P_l(\cos\theta) \tag{2.42}$$

We know that outside sphere, $A_l = 0$ and inside the sphere, $B_l = 0$. Therefore, potential within the sphere is given by (Fig. 2.3):

$$V(r,\theta) = \sum_{l=0}^{\infty}\left(A_l r^l\right)P_l(\cos\theta) \tag{2.43}$$

However, for the surface of the sphere, $r = a$. Therefore, Eq. (2.43) assumes the following form

$$V(\theta) = \sum_{l=0}^{\infty}\left(A_l a^l\right)P_l(\cos\theta) \tag{2.44}$$

$V(\theta)$ is the expansion with unknown coefficients $A_l a^l$. In order to determine A_l we have to consider inner product keeping in mind $x = \cos(\theta)$; $dx = -\sin(\theta)d\theta$

$$(P_{l'}, V(\theta)) = \sum_{l=0}^{\infty}\left(A_l a^l\right)(P_{l'}, P_l)$$

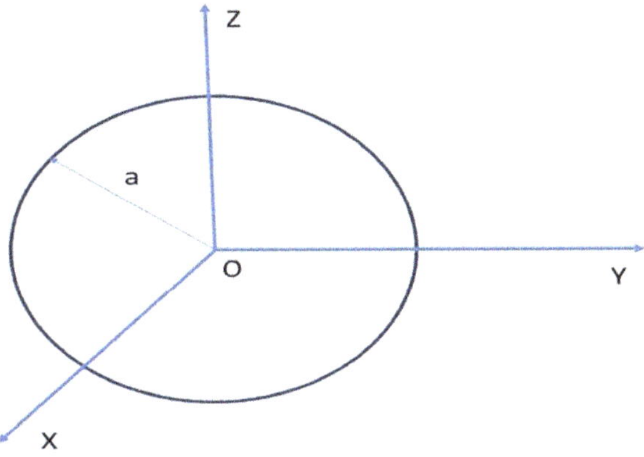

Fig. 2.3 A sphere of radius a

$$= \sum_{l=0}^{\infty} \left(A_l a^l\right) \frac{2}{2l+1} \delta_{l'l} \tag{2.45}$$

Therefore, we conclude that

$$A_l = \frac{2l+1}{2a^l} (P_l, V(\theta)) \tag{2.46}$$

Equation (2.46) can be illustrated in integral form as follows:

$$A_l = \frac{2l+1}{2a^l} \int_0^{\pi} d\theta \, \sin(\theta) P_l(\cos\theta) V(\theta) \tag{2.47}$$

On substituting value of A_l in $V(r, \theta) = \sum_{l=0}^{\infty} \left(A_l r^l\right) P_l(\cos\theta)$, we can get potential inside the sphere.

Example 2.2 Consider a spherical conductor with a radius a, designed from two interlocking hemispherical shells. The hemispheres are kept at equal and opposite potentials given by

$$V(\theta) = \begin{cases} +V \left(0 \leq \theta \leq \frac{\pi}{2}\right) \\ -V \left(\frac{\pi}{2} \leq \theta \leq \pi\right) \end{cases}$$

Prove that the potential with in the sphere is exhibited as follows:

$$V(r, \theta) = V \left[1 + \frac{3}{2} r/a P_1(\cos\theta) - \frac{7}{8} (r/a)^3 P_3(\cos\theta) \right. $$
$$\left. + \frac{11}{16} (r/a)^5 P_5(\cos\theta) + \cdots \right]$$

Here $P_l^{'s}(\cos\theta)$ are the Legendre polynomials.

Solution:

$$V(\theta) = \begin{cases} +V\left(0 \le \theta \le \frac{\pi}{2}\right) & (0 \le \cos\theta \le 1) \\ -V\left(\frac{\pi}{2} \le \theta \le \pi\right) & (-1 \le \cos\theta \le 0) \end{cases}$$

Inside the sphere, $B_l = 0$. Therefore, Eq. (2.47) can be written as follows (Fig. 2.4):

$$A_l = \frac{2l+1}{2a^l} \int_{-1}^{1} dx P_l(x) V(x) \tag{2.48}$$

$$A_l = \frac{2l+1}{2a^l} \left[\int_{0}^{1} dx P_l(x) V(x) + \int_{-1}^{0} dx P_l(x)(-V(x)) \right] \tag{2.49}$$

It is very laborious to solve this integral. Thus, to get A_l we use generating function. In order to consider the generating function, we examine the potential due to a point charge on the z-axis as viewed from another point. The fundamental expression for the potential in terms of distances r and r' is

$$V(r,\theta) = \sum_{l=0}^{\infty} \left(A_l r^l\right) P_l(\cos\theta) \tag{2.50}$$

We notice that only odd values of l will generate non-zero result.

Fig. 2.4 A spherical conducting shell of radius a

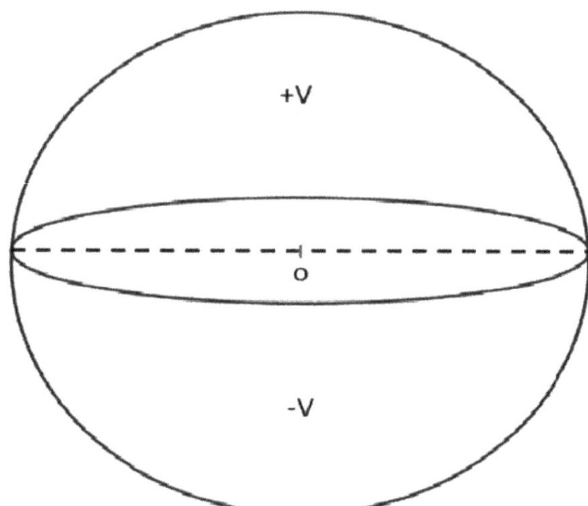

$P_0(x), P_2(x), P_4(x), \ldots\ldots$ will yield zero value. In another perspective, $P_1(x), P_3(x), P_5(x), \ldots\ldots$ will yield non-zero result. Let us substantiate our argument. However, we know that, $P_0(x) = 1$.

Pertinently, V is constant and, therefore, it can be pulled outside the integral in Eq. (2.49), we get

$$\int\limits_{0}^{1} dx - \int\limits_{-1}^{0} dx = 0$$

For $P_1(x) = x$, Eq. (2.49) becomes

$$\int\limits_{0}^{1} x dx - \int\limits_{-1}^{0} x dx = 2(1/2) = 1$$

So, for odd l values, we get A_l as follows:

$$A_l = \frac{2l+1}{2a^l} 2V \left[\int\limits_{0}^{1} dx P_l(x) \right] \tag{2.51}$$

If we put $P_l(x)$ from Rodrigue's formula as, $P_l(x) = \frac{1}{2^l l!} \left(\frac{d}{dx}\right)^l (x^2 - 1)^l$, Eq. (2.49) becomes

$$A_l = \frac{2l+1}{2a^l} 2V \left[\int\limits_{0}^{1} dx \frac{1}{2^l l!} \left(\frac{d}{dx}\right)^l (x^2 - 1)^l \right] \tag{2.52}$$

It is very laborious to solve this integral. Thus, to get A_l we use generating function. In order to consider the generating function, we examine the potential due to a point charge on the z-axis as viewed from another point. The fundamental expression for the potential in terms of distances r and r' is

$$\frac{1}{|\vec{r} - \vec{r'}|} = \frac{1}{\sqrt{r^2 + r'^2 - 2rr'\cos\theta}}$$

$$\frac{1}{|\vec{r} - \vec{r'}|} = \frac{1}{r\sqrt{1 + \left(\frac{r'}{r}\right)^2 - 2\frac{r'}{r}\cos\theta}}$$

We assume that $r' \leq r$; $\frac{r'}{r} \leq 1$ and put $\frac{r'}{r} = t, x = \cos\theta$. The generating function is defined as:

$$g(x, t) = \frac{1}{\sqrt{1 + (t)^2 - 2tx}} \tag{2.53}$$

The generating function can be enunciated in power series as follows:

$$g(x, t) = \sum_{n=0}^{\infty} P_n(x) t^n \tag{2.54}$$

And from which, P_n can be illustrated as under

$$P_n(x) = \frac{1}{n!} \frac{\partial^n}{\partial t^n} g(x, t)_{t \to 0} \tag{2.55}$$

We define,

$$I_l = I_n = \int_0^1 dx P_n(x) \tag{2.56}$$

$$I_G = \int_0^1 dx g(x, t) = \int_0^1 dx \frac{1}{\sqrt{1 + (t)^2 - 2tx}}$$

$$= \sum_{n=0}^{\infty} \int_0^1 dx P_n(x) t^n = \sum_{n=0}^{\infty} I_n t^n$$

$$\sum_{n=0}^{\infty} I_n t^n = I_0 t^0 + I_1 t^1 + I_2 t^2 + I_3 t^3 + I_4 t^4 + \cdots$$

$$I_G = \frac{-1}{t} \sqrt{1 + (t)^2 - 2tx} \Big|_0^1$$

$$I_G = \frac{-1}{t} \left[(1 - t) - \sqrt{1 + t^2} \right] \tag{2.57}$$

In order to solve the second term on the R.H.S of Eq. (2.57), we assume that, $x = t^2$. Therefore, we get

$$\sqrt{1 + x} = 1 + \frac{x}{2!} + \frac{1}{2!}(1/2)(-1/2)x^2 + \frac{1}{3!}(1/2)(-1/2)(-3/2)x^3 + \cdots$$

We can write above equation more explicitly as under:

$$\sqrt{1 + x} = 1 - \sum_{k=1}^{\infty} \frac{(2k - 3)!!}{(2k)!!}(-1)^k x^k \tag{2.58}$$

Substitute, Eq. (2.58) in Eq. (2.57), we get

$$I_G = \frac{-1}{t}\left[1 - t - 1 + \sum_{k=1}^{\infty} \frac{(2k-3)!!}{(2k)!!}(-1)^k t^{2k}\right] = \sum_{n=0}^{\infty} I_n t^n$$

$$I_G = 1 - \sum_{k=1}^{\infty} \frac{(2k-3)!!}{(2k)!!}(-1)^k t^{2k-1} \tag{2.59}$$

$$I_0 = 1;\ I_1 = 1/2;\ I_2 = 0;\ I_3 = -1/8;\ I_4 = 0.$$

We notice, that among even I_G only I_0, will contribute. Therefore, for $k > 0$, the general result is given by

$$I_{2k-1} = -(-1)^k \frac{(2k-3)!!}{2k!!}$$

where $k = 1, 2, 3\ldots$.

$$A_l = \frac{(2l+1)2VI_l}{2a^l} = \frac{(2l+1)VI_l}{a^l} \tag{2.60}$$

For, $l = 2k - 1$, Eq. (2.60) becomes

$$A_{l=2k-1} = \frac{V}{a^{2k-1}}[2(2k-1)+1](-1)^{k+1}\frac{(2k-3)!!}{2k!!} \tag{2.61}$$

For even values of l, only $l = 0$ will contribute, i.e., $I_0 = 1$
  Hence, we get $A_0 = V$.
  Therefore, the solution is given by

$$V(r,\theta) = \sum_{l=0}^{\infty}\left(A_l r^l\right)P_l(\cos\theta)$$

Here, $l = 2k - 1$, hence the above equation becomes

$$V(r,\theta) = V - V\sum_{k=1}^{\infty}(r/a)^{2k-1}(4k-1)\frac{(2k-3)!!}{2k!!}(-1)^k P_{2k-1}(\cos\theta) \tag{2.62}$$

Expanding first few terms, we get

$$V(r,\theta) = V\left[1 + \frac{3}{2}r/aP_1(\cos\theta) - \frac{7}{8}(r/a)^3 P_3(\cos\theta)\right.$$

$$\left. + \frac{11}{16}(r/a)^5 P_5(\cos\theta) + \cdots\right] \tag{2.63}$$

This is the solution of the sphere, with hemispheres at different potentials (Inside). In order to evaluate the potential just outside the sphere, $A_l = 0$, and $B_l \neq 0$. So, we can replace $(r/a)^l$ by $(a/r)^{l+1}$. In this way we can calculate potential outside the sphere.

Example 2.3 Let $V(\theta) = k \cos 3\theta$, where k is some constant, be the potential on the surface of the sphere of radius R. Deduce the following:

1. The potential inside the sphere ($r < R$).
2. The potential outside the sphere ($r > R$).
3. The surface charge density $\sigma(\theta)$ on the sphere.

Solution:

The solution to the Legendre differential equation with azimuthal symmetry is expressed by Eq. (2.42). Pertinently, $B_l = 0$, inside the sphere, therefore, Eq. (2.36) assumes the following form

$$V(r, \theta) = \sum_{l=0}^{\infty} \left(A_l r^l\right) P_l(\cos \theta) \tag{2.64}$$

However, outside the sphere, $A_l = 0$. Therefore, Eq. (2.38) becomes

$$V(r, \theta) = \sum_{l=0}^{\infty} \left(\frac{B_l}{r^{l+1}}\right) P_l(\cos \theta) \tag{2.65}$$

On the surface of the sphere

$$V(\theta) = k \cos 3\theta \tag{2.66}$$

We know that

$$\cos 3\theta = 4 \cos^3 \theta - 3 \cos \theta \tag{2.67}$$

We can write above equation in terms of Legendre polynomial as:

$$\cos 3\theta = a P_3(\cos \theta) + b P_1(\cos \theta)$$

$$\cos 3\theta = a \frac{\left(5 \cos^3 \theta - 3 \cos \theta\right)}{2} + b P_1(\cos \theta)$$

$$\cos 3\theta = \frac{5a \cos^3 \theta}{2} + \left(\frac{-3a \cos \theta}{2} + b \cos \theta\right) \tag{2.68}$$

Equating the coefficients of Eqs. (2.67) and (2.68), we get

$$\frac{5a}{2} = 4; \; a = 8/5$$

And

$$\frac{-3a}{2} + b = -3; b = -3/5$$

We could get the potential on the surface of the sphere by inserting Eq. (2.68) in Eq. (2.66), and hence, we obtain

$$V(\theta) = k \, \cos 3\theta = \frac{k}{5}(8P_3(\cos \theta) - 3P_1(\cos \theta)) \tag{2.69}$$

The solution on the surface of the sphere is written as follows:

$$V(r, \theta) = V(\theta) = \sum_{l=0}^{\infty} \left(A_l R^l\right) P_l(\cos \theta) \tag{2.70}$$

$$= \frac{k}{5}(8P_3(\cos \theta) - 3P_1(\cos \theta)) \tag{2.71}$$

We know from an orthonormal condition of Legendre polynomial

$$\int_0^\pi d\theta \, \sin(\theta) P_l(\cos \theta) P_{l'}(\cos \theta) = \frac{2}{2l + 1}$$

If we multiply Eq. (2.70) by $P_{l'}(\cos \theta) d\theta \, \sin(\theta)$ on both sides then the term $l' = l$ will survive

$$A_1 R(2/3) = \frac{k}{5}(-3) \int_0^\pi P_1(\cos \theta) P_1(\cos \theta) \sin(\theta) d\theta$$

$$A_1 R(2/3) = \frac{-3k}{5}(2/3)$$

$$A_1 = \frac{-3k}{5R} \tag{2.72}$$

Similarly, for A_3

$$A_3 R^3(2/7) = \frac{8k}{5} \int_0^\pi (P_3(\cos \theta))^2 \sin(\theta) d\theta$$

$$A_3 = \frac{8k}{5R^3} \tag{2.73}$$

$$V(r, \theta) = V(\theta) = \sum_{l=0}^{\infty} \left(A_l r^l\right) P_l(\cos \theta)$$

$$= \frac{-3k}{5R} r P_1(\cos \theta) + \frac{8k}{5R^3} r^3 P_3(\cos \theta)$$

$$= \frac{k}{5} \{8(r/R)^3 P_3(\cos \theta) - 3(r/R)P_1(\cos \theta)\}$$

$$\sigma(\theta) = \frac{\varepsilon_0 k}{5} \left(\frac{56}{R} P_3(\cos \theta) - \frac{9}{R} P_1(\cos \theta) \right) \tag{2.74}$$

On the surface

$$\sum_{l=0}^{\infty} \left(A_l R^l\right) P_l(\cos \theta) = \sum_{l=0}^{\infty} \frac{B_l}{R^{l+1}} P_l(\cos \theta)$$

$$\nabla^2 V = \frac{1}{r} \frac{\partial}{\partial r}\left(r \frac{\partial V}{\partial r} \right) + \frac{1}{r^2} \frac{\partial^2 V}{\partial \phi^2} + \frac{\partial^2 V}{\partial z^2} = 0 \tag{2.75}$$

$A_1, A_2 \rightarrow B_1, B_2$

$$B_1 = R^3 A_1 = R^3 \left(\frac{-3k}{5R} \right)$$

$$B_1 = -\frac{3kR^2}{5}$$

$$B_3 = A_3 R^7 = \left(\frac{8k}{5R^3} \right) R^7 = \frac{8k}{5} R^4$$

$$V(r, \theta) = \sum_{l=0}^{\infty} \frac{B_l}{r^{l+1}} P_l(\cos \theta) + \frac{B_3}{r^4} P_3(\cos \theta)$$

$$= \frac{-3k}{5} \frac{B_2}{r^2} P_1(\cos \theta) + \frac{8k}{5} \frac{R^4}{r^4} P_3(\cos \theta)$$

$$V_{\text{out}}(r, \theta) = \frac{k}{5} \left(8\left(\frac{R}{r}\right)^4 P_3(\cos \theta) - 3(R/r)^2 P_1(\cos \theta) \right) \tag{2.76}$$

In order to find surface charge density, we use boundary conditions

$$\frac{\partial V_{outside}}{\partial r} - \frac{\partial V_{inside}}{\partial r} = -\frac{\sigma(\theta)}{\varepsilon_0}$$

$$\frac{k}{5} \left(8R^4\left(-4r^{-5}\right) P_3(\cos \theta) - 3R^2\left(-2r^{-3} P_1(\cos \theta)\right) \right)$$

$$- \frac{k}{5} \left(8\left(\frac{3r^2}{R^3}\right) P_3(\cos \theta) - \frac{3}{R} P_1(\cos \theta) \right) = -\frac{\sigma(\theta)}{\varepsilon_0} \tag{2.77}$$

The surface charge density is given by inserting $r = R$ in above expression and solving it further we get

$$\sigma(\theta) = \frac{\varepsilon_0 k}{5}\left(\frac{56}{R}P_3(\cos\theta) - \frac{9}{R}P_1(\cos\theta)\right) \tag{2.78}$$

2.2.2 *Cylindrical Symmetry of Laplace Equation*

The Laplace equation in cylindrical coordinates is a critical mathematical framework used to describe physical systems with cylindrical symmetry, spanning applications in fluid mechanics, electromagnetism, acoustics, and heat transfer. Its solutions are fundamental to understanding various phenomena, including the distribution of electric potential, the behaviour of sound waves in cylindrical ducts, and the flow of fluids in pipes. The power of this equation lies in its ability to model situations where boundary conditions are imposed on cylindrical or circular geometries, which are common in both natural and engineered systems.

Writing Eq. (2.1) in cylindrical coordinates, we get

$$\nabla^2 V = \frac{1}{r}\frac{\partial}{\partial r}\left(r\frac{\partial V}{\partial r}\right) + \frac{1}{r^2}\frac{\partial^2 V}{\partial\phi^2} + \frac{\partial^2 V}{\partial z^2} = 0 \tag{2.79}$$

Here, r represents the radial coordinate, ϕ represents the azimuthal angle measured in the plane, and z indicates the vertical height along the z-axis. The importance of the cylindrical symmetry is that it removes Bessel functions from the solution of second-order differential equation. For symmetrical case where potential is not function of z, we can drop z-term. We employ the variable separable technique as follows:

$$V(r, \phi) = R(r)F(\phi) \tag{2.80}$$

Substitute Eq. (2.80) in Eq. (2.79) while dropping z-term in the later equation, we get

$$\frac{F}{r}\frac{d}{dr}\left(r\frac{dR}{dr}\right) + \frac{R}{r^2}\frac{d^2 F}{d\phi^2} = 0 \tag{2.81}$$

Each term in above equation is the function of (r, ϕ), and therefore, we multiply above equation by $\frac{r^2}{R(r)F(\phi)}$ throughout.

$$\frac{r}{R}\frac{d}{dr}\left(r\frac{dR}{dr}\right) + \frac{1}{F}\frac{d^2 F}{d\phi^2} = 0 \tag{2.82}$$

The first term in Eq. (2.82) depends on r alone; however, the second term exclusively depends on ϕ. We assume that the first and second terms are equal to n^2 and $-n^2$, respectively, except $n \neq 0$. Therefore, we write

$$\frac{r}{R}\frac{d}{dr}\left(r\frac{dR}{dr}\right) = n^2 \tag{2.83}$$

Also, we write

$$\frac{1}{F}\frac{d^2F}{d\phi^2} = -n^2 \tag{2.84}$$

Consider Eq. (2.83) and simplify it further, we get

$$\frac{r}{R}\frac{d}{dr}\left(r\frac{dR}{dr}\right) = n^2$$
$$r\left[rR'' + R'\right] = n^2R$$
$$\left[r^2R'' + rR'\right] - n^2R = 0 \tag{2.85}$$

We examine the series solution of this expression as follows:

$$R(r) = r^\lambda \tag{2.86}$$

Differentiate it with respect to λ ,we get

$$R'(r) = \lambda r^{\lambda-1} \tag{2.87}$$

Again differentiating it, we get

$$R''(r) = \lambda(\lambda - 1)r^{\lambda-2} \tag{2.88}$$

By inserting Eqs. (2.86), (2.87) and (2.88) into Eq. (2.85), we conclude

$$\lambda(\lambda - 1)r^\lambda + \lambda r^\lambda - n^2 r^\lambda = 0$$
$$\lambda^2 - n^2 = 0$$
$$\lambda = \pm n$$

Hence the solution of the r part is given by

$$R(r) = \left(Ar^n + \frac{B}{r^n}\right), \text{ were } n \neq 0 \tag{2.89}$$

And the solution of the ϕ part is given by:

$$F(\phi) = (C \, \sin(n\phi) + D \, \cos(n\phi)) \tag{2.90}$$

Therefore, the solution $V(r, \phi) = R(r)F(\phi)$

$$V(r, \phi) = \sum_{n=1}^{\infty} \left(Ar^n + \frac{B}{r^n} \right)(C \, \sin(n\phi) + D \, \cos(n\phi)) \tag{2.91}$$

This is the generalized solution of electric potential in cylindrical symmetry, while assuming that $n \neq 0$.

It is worthwhile to mention that $A = 0$ outside the cylinder and $B = 0$ inside it. Further, we check what happen for $n = 0$. From Eq. (2.82), we can write

$$\frac{r}{R}\frac{d}{dr}\left(r\frac{dR}{dr} \right) = 0$$
$$r\frac{dR}{dr} = a_1$$
$$\int dR = \int a_1 \frac{dr}{r}$$
$$R(r) = a_1 \log(r) + b_1 \tag{2.92}$$

Which is valid only for $n = 0$. Also, from Eq. (2.83), we can write

$$\frac{1}{F}\frac{d^2 F}{d\phi^2} = 0$$
$$\frac{dF}{d\phi} = c_1$$
$$F(\phi) = c_1\phi + d_1 \tag{2.93}$$

Which is also valid for $n = 0$. Thus, the solution is given by

$$V(r, \phi) = \left[a_1 \log(r) + b_1 \right][c_1\phi + d_1] \tag{2.94}$$

Hence, the general solution is given by

$$V(r, \phi) = \left(\left[a_1 \log(r) + b_1 \right][c_1\phi + d_1] \right)$$
$$+ \left(\sum_{n=1}^{\infty} \left(Ar^n + \frac{B}{r^n} \right)(C \, \sin(n\phi) + D \, \cos(n\phi)) \right) \tag{2.95}$$

where the constants can be ascertained from the initial boundary values. For instance, we consider the term

$$V(r, \phi) = a_1 \log(r) + b_1 \tag{2.96}$$

By definition as $r \to \infty$, $V \to 0$, which implies $a_1 = 0$.
So,

$$V(r, \phi) = b_1 \tag{2.97}$$

Further, we consider the term

$$V(r, \phi) = c_1 \phi + d_1 \tag{2.98}$$

where $0 \leq \phi \leq 2\pi$
   Potential should remain unchanged after $2n\pi$ rotations

$$V(r, \phi + 1000\pi) = c_1(\phi + 1000\pi) + d_1 \tag{2.99}$$

This shows symmetric potential is increasing, which is devoid of physical significance. Hence, we have to set $c_1 = 0$

$$V(r, \phi) = d_1 \tag{2.100}$$

Hence, we can write,

$$\left([a_1 \log(r) + b_1][c_1 \phi + d_1]\right) = b_1 d_1 = \text{constant} \tag{2.101}$$

$V(r, \phi) = \frac{a_0}{2}$, electric potential is constant rather than dependent on (r, ϕ).

$$V(r, \phi) = \left(\sum_{n=1}^{\infty} \left(Ar^n + \frac{B}{r^n}\right)(C \sin(n\phi) + D \cos(n\phi))\right)$$

As there are too many constants, we will redefine all these constants like $AC = A_n$; $AD = B_n$; $BC = C_n$; and $BD = D_n$;

$$V(r, \phi) = \sum_{n=1}^{\infty}[(A_n \sin(n\phi) + B_n \cos(n\phi))]r^n$$
$$+ \sum_{n=1}^{\infty}[(C_n \sin(n\phi) + D_n \cos(n\phi))]\frac{1}{r^n} \tag{2.102}$$

Let, a be the radius of the cylinder, we will also try to absorb the radius of the cylinder with in a constant, so we will redefine the constants.

$$V(r, \phi) = \sum_{n=1}^{\infty}[(a_n \sin(n\phi) + b_n \cos(n\phi))](r/a)^n$$

$$+ \sum_{n=1}^{\infty}[(c_n \sin(n\phi) + d_n \cos(n\phi))](a/r)^n \tag{2.103}$$

The ratio $(r/a)^n$ and $(a/r)^n$ decides the potential outside the cylinder and inside it, respectively. Together, with all the terms we can write the general solution of Laplace equation with cylindrical symmetry as: $d_n = -\frac{V_0}{n\phi}((-1)^n - 1)$

$$V(r, \phi) = \frac{a_0}{2} + \sum_{n=1}^{\infty}[(a_n \sin(n\phi) + b_n \cos(n\phi))](r/a)^n$$

$$+ \sum_{n=1}^{\infty}[(c_n \sin(n\phi) + d_n \cos(n\phi))](a/r)^n \tag{2.104}$$

provides a powerful framework for solving Laplace's equation in cylindrical geometries. The specific form of the potential depends on the boundary values. This solution is essential in problems involving charged conductors, boundary value problems and situations involving cylindrical symmetry.

Example 2.5 Consider a cylinder where the potential is defined as follows:

$$V(a, \phi) = \begin{cases} V_0; & 0 \le \phi \le \pi \\ 0; & \pi \le \phi \le 2\pi \end{cases}$$

Find the potential outside as well as inside the cylinder? (Fig. 2.5)

Solution:

(a) For outside cylinder $r > a$, the solution is given by

$$V(r, \phi) = \frac{a_0}{2} + \sum_{n=1}^{\infty}[(c_n \sin(n\phi) + d_n \cos(n\phi))](a/r)^n \tag{2.105}$$

Using the boundary conditions, we can determine Fourier coefficients

$$\frac{a_0}{2} = \frac{1}{2\pi}\int_0^{2\pi} V(a, \phi)\mathrm{d}\phi = \frac{1}{2\pi}\left[\int_0^{\pi} V_0 \mathrm{d}\phi + \int_{\pi}^{2\pi} 0 \mathrm{d}\phi\right] = \frac{V_0}{2} \tag{2.106}$$

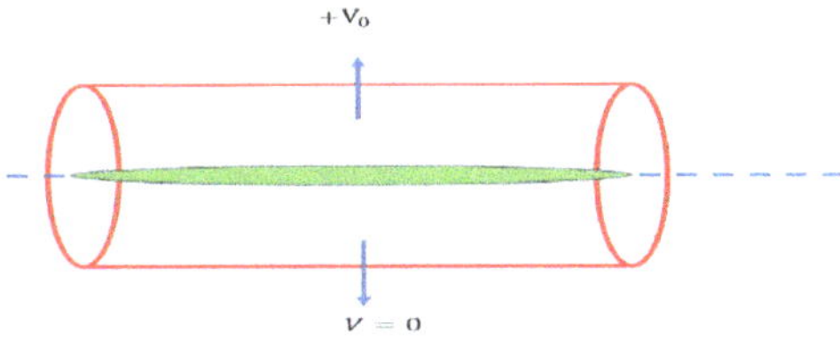

Fig. 2.5 A graphical representation of a cylindrical object with radius a, where the upper half is maintained at a positive potential and the lower half is set to zero potential

$$c_n = \frac{1}{\pi} \int_0^{2\pi} V(a, \phi)\,d\phi \cos(n\phi)$$

$$c_n = \frac{1}{\pi} \left[\int_0^{\pi} \cos(n\phi) V_0\,d\phi + \int_{\pi}^{2\pi} 0 \cdot \sin(n\phi)\,d\phi \right]$$

$$c_n = 0 \tag{2.107}$$

And

$$d_n = \frac{1}{\pi} \int_0^{2\pi} V(a, \phi)\,d\phi \sin(n\phi)$$

$$d_n = \frac{1}{\pi} \left[\int_0^{\pi} \sin(n\phi) V_0\,d\phi + \int_{\pi}^{2\pi} 0 \cdot \sin(n\phi)\,d\phi \right]$$

$$d_n = \frac{-V_0}{n\pi}\left((-1)^n - 1\right)$$

$$d_n = \begin{cases} \frac{2V_0}{n\pi}; & \text{if } n \text{ is odd} \\ 0; & \text{if } n \text{ is even} \end{cases} \tag{2.108}$$

Therefore, the electric potential outside the cylinder will be written as follows:

$$V(a, \phi) = \frac{V_0}{2} + \sum_{n=1,3\ldots}^{\infty} (a/r)^n \frac{2V_0}{n\pi} \cos(n\phi) \tag{2.109}$$

The dominant term in the summation is when $n = 1$

$$V(a, \phi) \approx \frac{V_0}{2} + \frac{2V_0 a \cos(\phi)}{r\pi} \tag{2.110}$$

(b)   To find $V(a, \phi)$ inside the cylinder, we can write:

$$V(a, \phi) \approx \frac{V_0}{2} + \sum_{n=1,3,\ldots}^{\infty} (r/a)^n \frac{2V_0}{n\pi} \sin(n\phi)$$

Dominant term is when $n = 1$ and is given by

$$V(a, \phi) \approx \frac{V_0}{2} + \frac{2V_0 r \sin(\phi)}{a\pi} \tag{2.111}$$

This is the procedure which is employed to get potential inside the cylinder.

2.3 Uniqueness Theorems and Boundary Constraints

The Laplace equation alone is insufficient to determine V explicitly, unless supplied by a suitable set of boundary conditions. Therefore, a question arises that what constitutes an appropriate set of boundary conditions that would help to determine V. In one dimensional case, we are usually confronted by just two constants in the general solution and therefore, we require only two boundary conditions. In this regard, we have several options for specifying boundary conditions: we can define the value of the function at both the ends, enumerate the value of the function and its derivative at one end, or we may fix the value of the function at one end and its derivative at the other. However, this information alone is rarely sufficient. It can be redundant if the conditions are identical, or inconsistent if they differ. In the case of two- and three-dimensional problems, we encounter partial differential equations, and identifying the correct boundary conditions is often a complex and tedious task. The adequacy of a proposed set of boundary conditions is typically established through a uniqueness theorem. In this discussion, we will focus on the two most crucial and widely applicable uniqueness theorems.

2.3.1 First Uniqueness Theorem

The solution to the Laplace equation within a specified region of volume $\mathcal{V}$ is uniquely determined if appropriate boundary conditions are provided on the surface enclosing $\mathcal{V}$. This principle is often formulated as the uniqueness theorem. In this context, we will explore the two most fundamental and useful versions of the theorem, which ensure that the solution to the Laplace equation is well-defined and singular under the given boundary constraints. These versions are critical in both theoretical and practical applications, as they guarantee that the solution does not depend on arbitrary or multiple possibilities when the boundary conditions are fixed (Fig. 2.6).

Proof We aim to demonstrate the uniqueness of the solution to the Poisson equation within a given volume $\mathcal{V}$, subject to either Dirichlet or Neumann boundary conditions on the closed surface S. To approach this, we assume, for the sake of contradiction, that there exist two distinct solutions, V_1 and V_2, which both satisfy the same boundary conditions, i.e.,

$$\nabla^2 V_1 = 0 \quad \text{and} \quad \nabla^2 V_2 = 0$$

Let

$$V_3 = V_2 - V_1$$

This will also obey Laplace equation

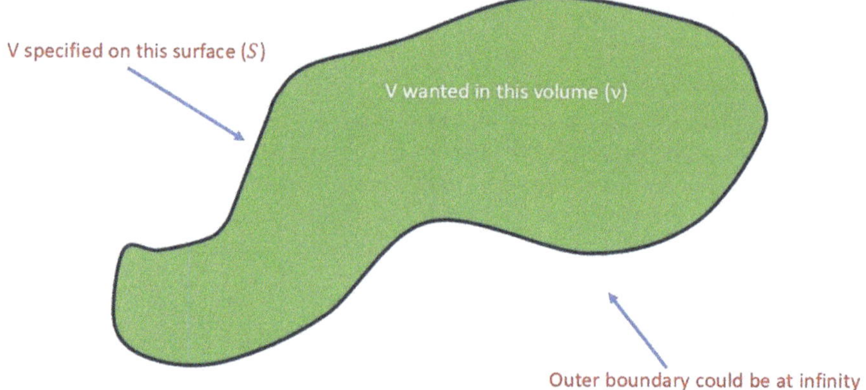

Fig. 2.6   A region of volume encompassing a surface S

$$\nabla^2 V_3 = \nabla^2 V_2 - \nabla^2 V_1$$

As already discussed, there is no local maxima and minima for potential

$$\nabla^2 V_3 = \nabla^2 V_2 - \nabla^2 V_1 = 0$$
$$V_1 = V_2$$

That verifies first uniqueness theorem. The specific method used to obtain the solution is irrelevant, provided that:

(a)   It satisfies the equation $\nabla^2 V = 0$
(b)   It adheres to the correct boundary conditions.

Now, $\nabla^2 V_1 = -\frac{\rho}{\varepsilon_0}; \ \nabla^2 V_2 = \frac{-\rho}{\varepsilon_0}$

$$\nabla^2 V_3 = \nabla^2 V_2 - \nabla^2 V_1 = -\frac{\rho}{\varepsilon_0} + \frac{\rho}{\varepsilon_0} = 0$$
$$V_3 = V_2 - V_1 = 0$$
$$V_1 = V_2$$

The potential in a volume $\mathcal{V}$ is uniquely determined if

(a)   Either the net charge density within the entire region is given or
(b)   Value of the potential V is specified on the boundaries.

2.3.2 Second Uniqueness Theorem

The simple procedure to obtain boundary conditions for the region of interest is to fix the potential V on all surfaces. However, there are situations where the potential at the boundary surface is unknown, but instead we are given charges on various conducting surfaces. Once a charge is placed on a conductor, it distributes itself according to certain rules. Therefore, it is more appropriate to enumerate the density of charge in a region between the conductors. Does this uniquely determine the electric field, or are there multiple ways by which the charges could arrange themselves on their respective conductors, each resulting in a different electric field? The electric field $\vec{E}$ within a certain volume $\mathcal{V}$, encompassed by conductors and possessing a specified charge density ρ, is distinctively governed if the total charge on each conductor is known. This region can either be enclosed by another conductor or extend unbounded.

Proof The region between the conductors contains charge density ρ; therefore, we can apply Gauss's law as follows:

$$\vec{\nabla} \cdot \vec{E}_1 = \frac{\rho}{\varepsilon_0}; \vec{\nabla} \cdot \vec{E}_2 = \frac{\rho}{\varepsilon_0}$$

For ith conducting surface, we can write (Fig. 2.7)

$$\oint \vec{E}_1 \cdot \mathrm{d}\vec{a} = \frac{Q_i}{\varepsilon_0}; \oint \vec{E}_2 \cdot \mathrm{d}\vec{a} = \frac{Q_i}{\varepsilon_0}$$

Likewise, on the outer boundary we have charge $= Q_{\text{total}}$

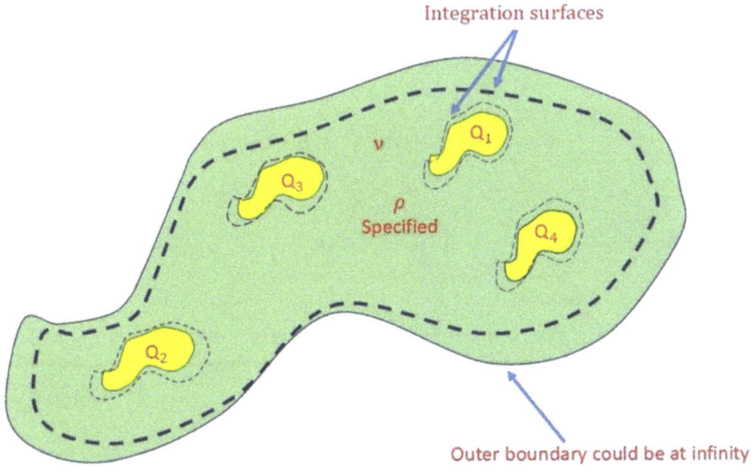

Fig. 2.7 Graphical representation of a conductor having charge density ρ

$$\oint \vec{E}_1 \cdot d\vec{a} = \frac{Q_{\text{total}}}{\varepsilon_0}; \quad \oint \vec{E}_2 \cdot d\vec{a} = \frac{Q_{\text{total}}}{\varepsilon_0}$$

$$\vec{E}_3 = \vec{E}_1 - \vec{E}_2$$

$$\vec{\nabla} \cdot \vec{E}_3 = \vec{\nabla} \cdot \vec{E}_1 - \vec{\nabla} \cdot \vec{E}_2$$

$$= \frac{\rho}{\varepsilon_0} - \frac{\rho}{\varepsilon_0} = 0$$

$$\vec{\nabla} \cdot \vec{E}_3 = 0$$

$$\oint \vec{E}_3 \cdot d\vec{a} = 0$$

However, the potential across the entire conductor is uniform. Therefore, every conducting surface is equipotential. Thus, we may write

$$\vec{\nabla} \cdot \left(V_3 \vec{E}_3 \right) = V_3 \left(\vec{\nabla} \cdot \vec{E}_3 \right) + \vec{E}_3 \cdot \left(\vec{\nabla} V_3 \right)$$

$$= 0 + \vec{E}_3 \cdot \left(-\vec{E}_3 \right)$$

$$= -(E_3)^2$$

Integrating over volume

$$\oint_\tau \vec{\nabla} \cdot \left(V_3 \vec{E}_3 \right) d\tau = \oint_S V_3 \cdot \vec{E}_3 d\vec{a}$$

$$= - \oint_\tau (E_3)^2 d\tau$$

where V_3 is constant over the boundary

$$0 = V_3 \oint_S \vec{E}_3 \cdot d\vec{a} = - \oint_\tau (E_3)^2 d\tau$$

$$\oint_\tau (E_3)^2 d\tau = 0$$

The integrand is always non-negative, and the only possibility is $E_3 = 0$

$$\vec{E}_1 = \vec{E}_2$$

Example 2.5 Imagine a conducting spherical shell with an outer radius R, which is electrically grounded, meaning it is held at a constant potential of $V = 0$. Inside this shell, at its exact geometric centre, a point charge q is placed. Use the first uniqueness theorem to argue that the potential inside the shell is uniquely determined by these boundary conditions.

Solution:

The cylinder is grounded, therefore, $V = 0$ at $r = R$, are the requisite boundary constraints.

At the centre, the point charge q induces a singularity at $r = 0$, but we know that $V(r) \to 0$ as $r \to 0$.

Laplace's Equation: Since there are no charges within the spherical shell (other than the point charge at the centre, which we'll treat separately), Laplace's equation is satisfied by the potential:

$$\nabla^2 V = 0 \quad \text{For } 0 < r < R$$

First Uniqueness Theorem: According to it, the solution of Laplace's equation is uniquely determined if potential is fixed on the boundary (in this case, $V = 0$ at $r = R$). Since these conditions are met, the potential inside the shell is uniquely determined.

Consider a point charge q located at the centre of a spherical shell that is electrically grounded, we know from symmetry that the potential must depend only on r, and for the region $0 < r < R$, the solution becomes

$$V(r) = \frac{q}{4\pi \varepsilon_0 r}$$

Thus, the potential is uniquely specified based on the given boundary conditions and inherent symmetry.

Example 2.6 A long grounded cylindrical conductor of radius R is placed along z-axis. Inside the cylinder, a line charge per unit length is placed at a distance a from the axis, where $a < R$. Use the second uniqueness theorem that electric field is uniquely determined by the boundary conditions.

Solution:

Boundary conditions: The cylinder is grounded, $V = 0$ at $r = R$.

The line charge λ creates symmetry and the potential V should reflect the cylindrical geometry.

Poisson's equation: In the region inside the cylinder where there is a line charge, Poisson's equation governs the potential.

$$\nabla^2 V = -\frac{\lambda}{\varepsilon_0} \delta(r - a)\delta(\theta - \theta_0)$$

where r is the radial distance from the cylinder axis, and θ_0 is the angular position of the line charge.

Second Uniqueness Theorem: According to it, the solution of Poisson's equation within a conductor is unique if the charge distribution and the potential on the boundary are specified. Here, the potential, $V = 0$ when, $r = R$. The charge distribution is given by the line charge at $r = a$. By invoking the second uniqueness theorem, we deduce that the electric field and potential inside the cylinder are uniquely determined by these boundary conditions.

Unsolved Problems:

Problem 2.1 Consider a spherical shell of radius R, with no charge inside. On the shell's surface, the potential is expressed as

$$V(R, \theta) = V_0 \cos^2 \theta$$

where V_0 is a constant. Find the potential $V(r, \theta)$ inside the shell for $r < R$?

Ans. $V(r, \theta) = \frac{V_0}{3} + \frac{V_0}{3}\left(\frac{r^2}{R^2}\right)(3\cos^2 \theta - 1)$.

Problem 2.2 A conducting sphere of radius R possesses a surface charge distribution characterized by $\sigma(\theta) = \sigma_0 \cos \theta$, where, σ_0 is a constant. Determine the electric potential $V(r, \theta)$ in the region outside the sphere $r > R$ due to this non-uniform surface charge distribution?

Ans. $V(r, \theta) = \frac{\sigma_0 R^2}{\varepsilon_0 r^2} \cos \theta$.

Problem 2.3 The potential due to a spherical shell of radius R is $V(r, \theta) = V_0 \cos \theta$ on its surface, where, V_0 is a constant. Find the electrostatic potential $V(r, \theta)$ inside the shell using azimuthal symmetry?

Ans. $V(r, \theta) = \frac{V_0}{R} r \cos \theta$.

Problem 2.4 Find a charge distribution that would produce the Yukawa potential
$$\phi = \frac{q}{4\pi \varepsilon_0} \frac{e^{-r/a}}{r}$$
Ans. $\rho = -\frac{q}{4\pi a^2} \frac{e^{-r/a}}{r}$.

Problem 2.5 A square sheet is charged uniformly with charge density σ. Show that the potential at the centre of the square is $\phi = \frac{\sigma a}{\pi \varepsilon_0} \ln(1 + \sqrt{2})$, where a is the length of the side of the square.

Problem 2.6 Consider a rectangular metal box with sides a, b and c along the x, y, and z axes, respectively. The box is grounded on three sides, while the fourth side (at $x = a$) is held on an electrical potential $V(x = a, y, z) = V_0(y, z)$. Show that the potential inside the box is uniquely determined using the first uniqueness theorem.

Problem 2.7 In a coaxial cable the potential of the outer cylinder of radius b is maintained at zero and that of the inner cylinder of radius a is V_1. Find the expression for the potential at a point in the region between two cylinders.

Ans. $V(r) = -\frac{V_1}{\ln (b/a)} \ln (r/b)$.

Problem 2.8 Solving Poisson's equation $\nabla^2\phi = -\frac{\rho_0}{\varepsilon_0}$ for the electrostatic potential $\phi(\vec{x})$ in a region with charge density ρ_0, two students find different solutions, viz. $\phi_1(\vec{x}) = -\frac{1}{2}\frac{\rho_0}{\varepsilon_0}x^2$ and $\phi_2(\vec{x}) = -\frac{1}{2}\rho_0\varepsilon_0 y^2$. Discuss the reason why both of these different solutions are correct.

Problem 2.9 A line charge of linear density λ is placed at a distance d above an infinite grounded conducting plane. Find the potential in the region above the plane.

Ans. $V(r) = \frac{\lambda}{2\pi\varepsilon_0}\ln\left(\frac{r_1}{r_2}\right)$.

Problem 2.10 Consider an infinitely long cylinder of radius R, placed along the z-axis, which carries a static charge density $\rho(\vec{r}) = kr$, where r is the perpendicular distance from the axis of the cylinder and k is a constant. The electrostatic potential inside the cylinder is $\phi(\vec{r}) \propto \left(\frac{r^3}{R^3} - 1\right)$.

2.4 Summary

- **The Laplace Equation**: The Laplace equation $\nabla^2 V = 0$ is emphasized as a core equation in electrostatics, widely applicable in electrodynamics for regions with no charge density. We explored its solutions in spherical coordinates.
- **Separation of Variables**: The separation of variable technique is used to simplify the Laplace equation by breaking it into parts based on spherical coordinates, yielding solutions that depend on the radial, polar and azimuthal angles.
- **Legendre Polynomials and Azimuthal Symmetry**: In this chapter we discuss the Legendre differential equation and Legendre polynomials, which arise naturally in problems with spherical symmetry. Rodrigue's formula provides a systematic way to derive these polynomials.
- **Boundary Conditions in Spherical and Cylindrical Systems**: We have studied the boundary conditions and their importance in determining potential:
- **Spherical Symmetry**: Solutions involve finding potential inside or outside a spherical conductor with specific boundary conditions.
- **Cylindrical Symmetry**: The Laplace equation in cylindrical coordinates, useful for systems with cylindrical geometries, like wires or pipes. Solutions often involve Bessel functions.
- **Application of Laplace's Equation to Physical Problems**: Several examples illustrate applying the Laplace equation in spherical and cylindrical systems, showing how to derive potential in systems with specified boundary values or symmetry.
- **Generating Function and Recursion Relations for Legendre Polynomials**: The chapter provides methods for calculating Legendre polynomials systematically using generating functions and recursion relations, essential in boundary value problems involving spherical symmetry.

- **Electrostatic Potential for Specific Configurations**: Examples include calculating potential inside or outside conductors with given boundary conditions (e.g., potential on a spherical shell or cylindrical conductor), reinforcing the application of these methods.
- **First Uniqueness Theorem**: This theorem guarantees a unique solution to the Laplace equation if the potential on a boundary is known, ensuring that no arbitrary solutions exist.
- **Second Uniqueness Theorem**: This theorem asserts that the electric field within a volume is uniquely determined if the charge distribution and conductor charges on the boundary are specified.

Chapter 3
Boundary Value Problems-II

Abstract This chapter presents a comprehensive exploration of methods and concepts in electrostatics and electrodynamics, focusing on three primary techniques: the separation of variables, the method of images and the Finite Element Analysis for two-dimensional cases. The method of images is introduced as a powerful tool for solving boundary problems by strategically placing imaginary charges outside the region of interest to satisfy boundary conditions. Detailed derivations are provided for the potential, electric field, surface charge density, Coulomb force and the work required to move charges in various configurations. A significant portion of the chapter is devoted to understanding the interaction between a point charge and a grounded conducting sphere. This analysis includes deriving the potential and force due to image charges, as well as examining special cases where the force approximates Coulomb's law at short distances and deviates at long distances. The induced surface charge density on the sphere is also derived and analyzed for different angular positions. Further, the interaction between a point charge and an insulated conducting sphere is explored, incorporating the superposition principle to determine the resulting potential and forces. The behaviour of a conducting sphere in a uniform electric field is also examined, highlighting how image charges satisfy boundary conditions and how the induced surface charge density varies across the sphere's surface. The chapter extends to advanced topics such as multipole expansion, which is used to describe the electric potential of complex charge distributions. This includes the monopole, dipole and quadrupole contributions, with particular emphasis on the behaviour of these potentials at large distances. The concept of vector potential is introduced, using multipole moments to analyze magnetic fields and the magnetic dipole moment, emphasizing its consistency with Maxwell's equations. Theoretical developments are complemented by applications that include the calculation of potentials, forces and surface charge densities for dipoles and quadrupoles, as well as the analysis of magnetic field configurations. To reinforce the concepts, the chapter includes solved examples and unsolved problems that encourage further exploration and practical application of the principles discussed.

Keywords Potentials · Method of images · Multipole expansion · Boundary conditions

R. Ahmad Parra et al., *A Systematic Approach to Electrodynamics*,
https://doi.org/10.1007/978-981-96-3408-8_3

3.1   Introduction

The following three methods are generally employed to solve the Laplace or Poisson's equations:

(a)   Method of Variable Separation
(b)   Method of Images
(c)   Finite Element Analysis (limited to 2D cases).

The technique, separation of variables, has already been discussed in the previous chapter. In the present context, we will discern problems based on the method of images.

The method of images addresses the challenge of handling one or more-point charges near boundary surfaces. When certain conditions are met, the geometry of the setup can suggest that a limited number of strategically placed charges, with carefully chosen magnitudes, can effectively replicate the requisite boundary values. These substitute charges are dubbed as image charges and such charges are utilized to discuss the original problem. The technique often referred as the method of images involves using imaginary or "mirror" charges in an extended region to satisfy boundary conditions in electrostatics. The image charges are generally positioned outside the domain of interest. In order to explain this, let us consider the example of potential on the plane (Fig. 3.1).

$$V(\vec{x}) = \frac{Kq}{|\vec{x} - \vec{x}'|} - \frac{Kq}{|\vec{x} - \vec{x}''|} \qquad (3.1)$$

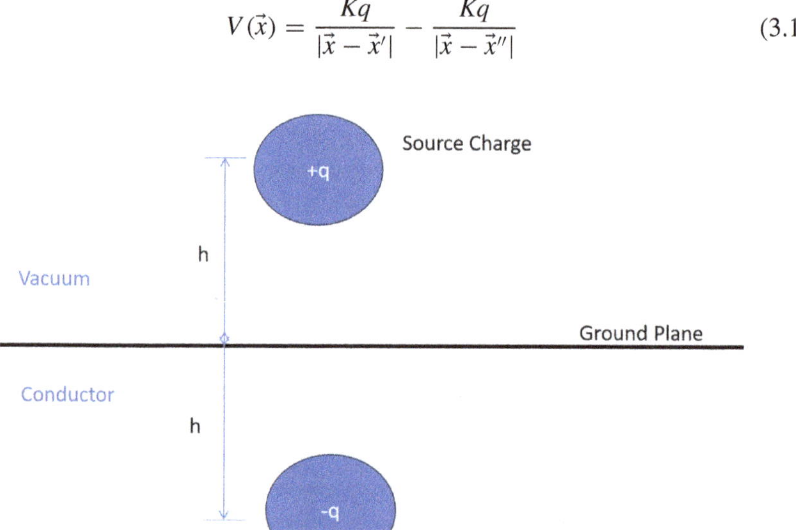

Fig. 3.1 Schematic representation of image charges

where $\vec{x}'$ represents the position of source charge q and $\vec{x}''$ represents the locations of image charge $-q$. From this we can deduce the following quantities.

(1) **Electric Field**: This is generally expressed as the negative gradient of the scalar potential V, and is therefore, mathematically written as follows

$$\vec{E} = -\vec{\nabla}V = -\frac{\partial V}{\partial n}\hat{n} \tag{3.2}$$

(2) **Charge Density**: The surface charge density can be determined by taking the normal derivative of the potential V at the boundary surfaces.

$$\sigma = -\frac{1}{4\pi K}\frac{\partial V}{\partial n} \tag{3.3}$$

(3) **Coulomb Force**: The interaction between a source charge and an image charge is defined as under

$$F = \frac{Kq_1q_2}{r^2} = Kq^2/(2h)^2 \tag{3.4}$$

(4) **Work**: The energy dissipated to get a charge at a localized position from infinity within a specific electric potential produced by a charge q, can be obtained as follows:

$$W = qV \tag{3.5}$$

The above expression can be further simplified as follows:

$$W = \frac{Kq^2}{2h} \tag{3.6}$$

3.2 Electrostatic Interaction of a Point Charge with a Grounded Conducting Sphere

The conducting sphere implies that if any charge is kept on one point of the surface of the sphere, its effect will be perceived throughout the entire spherical surface. Here, q is the source charge and q' is the image charge as shown in Fig 3.2.

Potential: The potential at point P may be expressed as:

$$V(\vec{x}) = \frac{Kq}{|\vec{x} - \vec{y}|} + \frac{Kq'}{|\vec{x} - \vec{y}'|} \tag{3.7}$$

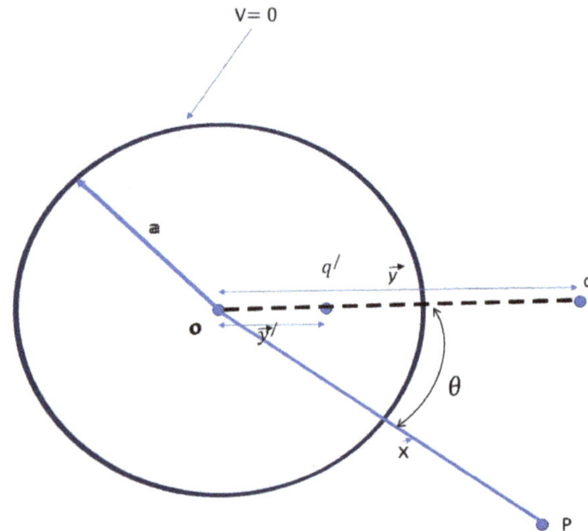

Fig. 3.2 Conducting sphere with radius 'a' possessing charge q produces an induced image charge q'. This arrangement guarantees the appropriate distribution of the electric field, fulfilling the necessary boundary values at the sphere's surface

The potential on the surface of the sphere should vanish, i.e.; $V(\vec{x})|_{x=a} = 0$, which implies that

$$\frac{Kq}{|\vec{x} - \vec{y}|} = -\frac{Kq'}{|\vec{x} - \vec{y}'|}$$

$$\frac{q'}{q} = -\frac{|\vec{x} - \vec{y}'|}{|\vec{x} - \vec{y}|}$$

$$\left(\frac{q'}{q}\right)^2 = \left(-\frac{|\vec{x} - \vec{y}'|}{|\vec{x} - \vec{y}|}\right)^2 = \frac{\vec{x}.\vec{x} + \vec{y}'.\vec{y}' - 2\vec{x}.\vec{y}'}{\vec{x}.\vec{x} + \vec{y}.\vec{y} - 2\vec{x}.\vec{y}} \tag{3.8}$$

The boundary value condition implies that $V(\vec{x})|_{x=a} = 0$

$$\left(\frac{q'}{q}\right)^2 \times \left(a^2 + y^2 - 2ay\cos(\theta)\right) - \left(a^2 + y'^2 - 2ay'\cos(\theta)\right) = 0$$

Rearranging the terms in above equation, we get (Fig. 3.2)

$$\left[\left(\frac{q'}{q}\right)^2(-2ay) + 2ay'\right]\cos(\theta) + \left(\frac{q'}{q}\right)^2\left(a^2 + y^2\right) - \left(a^2 + y'^2\right) = 0 \tag{3.9}$$

From the above equation we can write,

$$\left(\frac{q'}{q}\right)^2(-2ay) + 2ay' = 0$$

$$\left(\frac{q'}{q}\right)^2 = \frac{y'}{y}$$

$$q' = \pm q\sqrt{\frac{y'}{y}} \tag{3.10}$$

This represents the induced charge in terms of known charge. One root is significant from the geometry, as physically q should be $-q$. From the 2nd term of Eq. (3.9), we can write

$$\left(\frac{q'}{q}\right)^2 \left(a^2 + y^2\right) - \left(a^2 + y'^2\right) = 0$$

$$\left(\frac{y'}{y}\right)\left(a^2 + y^2\right) - \left(a^2 + y'^2\right) = 0$$

Multiply the above equation by y throughout, we get:

$$y'\left(a^2 + y^2\right) - y\left(a^2 + y'^2\right) = 0$$

$$y'^2 - \left(\frac{a^2 + y^2}{y}\right)y' + a^2 = 0$$

$$\left(y' - \frac{a^2}{y}\right)\left(y' - y\right) = 0 \tag{3.11}$$

which is quadratic equation in y'. This will yield either $y' = y$, or $y' = \frac{a^2}{y}$. The first solution is of no use, however, from second solution, we can express y' in guise of a and y, which implies $q' = -q\frac{a}{y}$ and $y' = \frac{a^2}{y}$. From this we can infer that we can fix the potential on the boundary by just adjusting y. Furthermore, we can conclude that the charges reside on the surface and there are no electric charges present within its interior.

Force of Interaction: The interaction between charges q and q' is enumerated as follows:

$$F = \frac{Kqq'}{(y - y')^2} \tag{3.12}$$

$$F = \frac{-Kq^2\frac{a}{y}}{\left(y - \frac{a^2}{y}\right)^2} = \frac{K\left(\frac{-q^2}{a^2}\right)}{\left(\frac{y}{a}\right)^3\left(1 - \frac{a^2}{y^2}\right)^2}$$

$$F = -\frac{K\frac{q^2}{a^2}}{\left(\frac{y}{a}\right)^3\left(1 - \frac{a^2}{y^2}\right)^2} \tag{3.13}$$

We will discuss some special cases as follows:

(a) **Short Distance**: In this case $y \approx a$, and from Eq. (3.13) we can write

$$F \approx -\frac{Kq^2}{a^2}\left(1 - \frac{a^2}{y^2}\right)^{-2}$$

Using binomial expansion $(1 - x)^n \approx 1 - nx$; where we have neglected high order terms

$$F \approx -\frac{Kq^2}{a^2}\left(1 + \frac{2a^2}{y^2}\right) \approx \frac{1}{y^2} \qquad (3.14)$$

The above equation resembles with the Coulomb's law.

(b) **Long Distance**: In this case $y \gg a$, and from Eq. (3.13), we obtain

$$F \approx -\frac{Kq^2}{a^2}\frac{a^3}{y^3} \qquad (3.15)$$

For considerable distances between charges, the force diminishes in proportion to the cube of the separation. This principle reflects the modified form of Coulomb's law, i.e., Coulomb's law is not true when we go very far away, it is not a universal law.

Induced Surface Charge Density

It has been previously mentioned that the induced surface charge density can be determined by calculating the normal derivative of the potential at that surface. Therefore, from Eq. (3.7) we get

$$\sigma = -\frac{1}{4\pi K}\frac{\partial V}{\partial x}\bigg|_{x=a}$$

$$\sigma = -\frac{1}{4\pi K}\frac{\partial}{\partial x}\left(\frac{Kq}{\left(x^2 + y^2 - 2xy\mathrm{Cos}(\theta)\right)^{1/2}} + \frac{Kq'}{\left(x^2 + y'^2 - 2xy'\cos(\theta)\right)^{1/2}}\right)_{x=a}$$

At $x = a$, that is on the surface of the sphere

$$\sigma = \frac{1}{4\pi}\left(\frac{1}{2}\frac{q(2a - 2y\cos(\theta))}{\left(a^2 + y^2 - 2ay\cos(\theta)\right)^{3/2}} + \frac{1}{2}\frac{q'(2a - 2y'\cos(\theta))}{\left(a^2 + y'^2 - 2ay'\cos(\theta)\right)^{3/2}}\right)$$

Using, $q' = -q\frac{a}{y}$, $y' = \frac{a^2}{y}$, in the above equation, we get

$$\sigma = \frac{1}{4\pi}\left(\frac{q(a - y\cos(\theta))}{\left(a^2 + y^2 - 2ay\cos(\theta)\right)^{3/2}} + \frac{-q\frac{a}{y}\left(a - \frac{a^2}{y}\cos(\theta)\right)}{\left(a^2 + \frac{a^4}{y^2} - 2a\frac{a^2}{y}\cos(\theta)\right)^{3/2}}\right)$$

$$\sigma = \frac{1}{4\pi}\left(\frac{q(a - y\cos(\theta))}{y^3\left(\frac{a^2}{y^2} + 1 - 2\frac{a}{y}\cos(\theta)\right)^{3/2}} - \frac{q\left(a - \frac{a^2}{y}\cos(\theta)\right)}{a^2 y\left(1 + \frac{a^2}{y^2} - 2\frac{a}{y}\cos(\theta)\right)^{3/2}}\right)$$

$$\sigma = \frac{1}{4\pi}\frac{q}{\left(1 + \frac{a^2}{y^2} - 2\frac{a}{y}\cos(\theta)\right)^{3/2}} \times \left(\frac{1}{y^3}(a - y\mathrm{Cos}(\theta)) - \frac{1}{a^2 y}\left(a - \frac{a^2}{y}\cos(\theta)\right)\right)$$

$$\sigma = \frac{q}{4\pi}\frac{1}{\left(1 + \frac{a^2}{y^2} - 2\frac{a}{y}\cos(\theta)\right)^{3/2}} \times \left(\frac{a}{y^3} - \frac{1}{ay}\right)$$

The above equation can be simplified as under:

$$\sigma = -\frac{q}{4\pi a^2}\left(\frac{a}{y}\right)\frac{\left(1 - \frac{a^2}{y^2}\right)}{\left(1 + \frac{a^2}{y^2} - 2\frac{a}{y}\cos(\theta)\right)^{3/2}} \tag{3.16}$$

The quantity $\frac{q}{4\pi a^2}$ in the above equation has the dimensions of surface charge density. Therefore, we can say that the quantity σ represents the normalized surface charge density. The dimensionless quantities, viz.; $\frac{a}{y}$ and $\frac{y}{a}$ are called generators. However, the quantity, $\frac{y}{a}$ exhibits how far we have moved from the surface of the sphere.

For example, if $\frac{y}{a} = 2$, i.e., $y = 2a$, that means we are going far way two times the radius of the sphere, similarly, if $\frac{y}{a} = 4$, i.e., $y = 4a$, that means we are going far way four times the radius of the sphere here we define $\sigma' = \frac{q}{4\pi a^2}$

$$\sigma = -\sigma'\left(\frac{a}{y}\right)\frac{\left(1 - \frac{a^2}{y^2}\right)}{\left(1 + \frac{a^2}{y^2} - 2\frac{a}{y}\cos(\theta)\right)^{3/2}} \tag{3.17}$$

From the above expression we are going to deduce some of the elegant physical concepts. As we know that θ is a dimensionless quantity, σ should also be dimensionless quantity. We will plot graph between these two dimensionless quantities.

For example,

$$y = 2a$$

$$-\frac{4\pi a^2 \sigma}{q} = \frac{1}{2}\frac{3/4}{\left(\frac{5}{4} - \cos(\theta)\right)^{3/2}} \tag{3.18}$$

For $\theta = 0^0$

$$-\frac{4\pi a^2 \sigma}{q} = 3$$

For $\theta = \frac{\pi}{2}$

$$-\frac{4\pi a^2 \sigma}{q} = 0.4$$

For $\theta = \pi$

$$-\frac{4\pi a^2 \sigma}{q} = 0.1$$

From the above discussion, it is evident that $as\theta$ increases from 0 to $\frac{\pi}{2}$, the normal charge density decreases rapidly; however, it decreases further slowly with the increase in θ. It looks like we are moving away from the point where surface charge density due to q' decreases by changing the angle θ (Fig. 3.3).

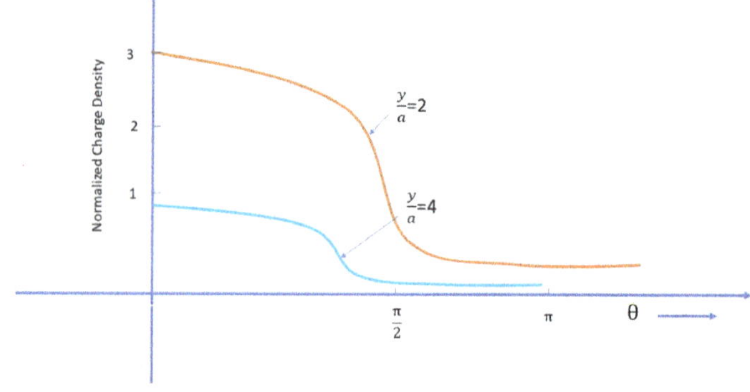

Fig. 3.3 The schematic representation of surface charge density of a conducting sphere having radius 'a'' generated due to the presence of a point charge situated at a distance y from the sphere's centre. Note that the scale in the graph is not precisely accurate and is for illustrative purposes only

3.3 Point Charge (q) Near a Charged (Q) Insulated Conducting Sphere

As discussed in the previous section, we are aware of the fact that the image charges inhabit the surface of a conductor. But for practical purposes, we assume it to be concentrated at the origin of a sphere. Since, we have proved in case of a point charge q near a grounded sphere, the induced charge on the surface of a sphere is given by $q' = -q\frac{a}{y}$.

In addition, let us examine an insulated conducting sphere that possesses a total charge Q in presence of a point charge q. We can approach the potential function for this scenario using the principle of linear superposition. First, we define the conducting sphere as being grounded, with a charge q' distributed uniformly throughout its entire surface. After grounding the sphere, we disconnect the ground wire and add an additional charge of $(Q - q')$ to the sphere. For the sake of simplicity, we will assume this additional charge is also uniformly distributed over the surface of the insulated conducting sphere. It is important to note that the electrostatic forces exerted by the point charge q are counterbalanced by the charge q' on the sphere (Fig. 3.4).

Let

$$q'' = Q - q' \text{(total charge } q'' + q') \tag{3.19}$$

$$q'' = Q + q\frac{a}{y} \tag{3.20}$$

We also know that $y' = \frac{a^2}{y}$.

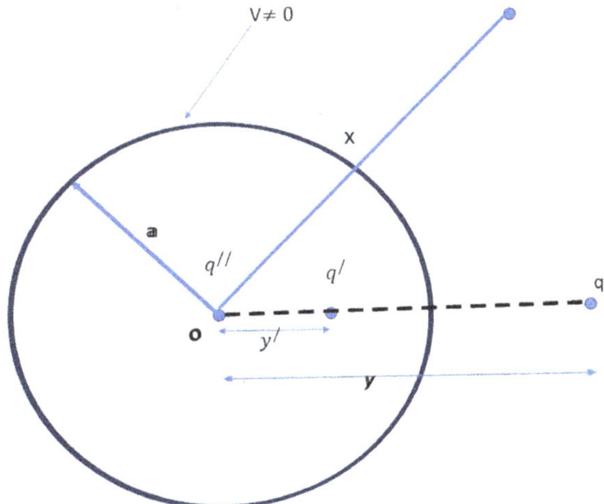

Fig. 3.4 A point charge q near an insulated conducting sphere of charge Q

Potential

The electric potential at an arbitrary point x, located outside the sphere, is influenced by the charges q, q', and q'' is, therefore, according to the superposition principle

$$V(\vec{x}) = \frac{Kq}{|\vec{x} - \vec{y}|} + \frac{Kq'}{|\vec{x} - \vec{y}'|} + \frac{q''}{|\vec{x}|}$$

$$V(\vec{x}) = K\left(\frac{q}{|\vec{x} - \vec{y}|} - \frac{q\frac{a}{y}}{\left|\vec{x} - \frac{a^2}{y^2}\vec{y}\right|} + \frac{Q + q\frac{a}{y}}{|\vec{x}|}\right) \tag{3.21}$$

This is the potential due to three charges. It differs from the previous case by the fact that the sphere under consideration is not grounded, however, it is insulated and, therefore, the charges does not flow anywhere. The additional surface charge density is contributed by the charge q'' and is equal to $\frac{q''}{4\pi a^2}$. Therefore, the total surface charge density is given by

$$\sigma = \frac{q}{4\pi a^2} \frac{\left(\frac{a}{y}\right)\left(\frac{a^2}{y^2} - 1\right)}{\left(1 + \frac{a^2}{y^2} - 2\frac{a}{y}\cos(\theta)\right)^{3/2}} + \frac{Q + q\frac{a}{y}}{4\pi a^2} \tag{3.22}$$

Force of Interaction: The interaction experienced by charge q due to the presence of charges q' and q'' is expressed by the following equation

$$F_q = \frac{Kqq''}{y^2} + \frac{Kqq'}{|\vec{y} - \vec{y}'|^2}$$

$$\vec{F}_q = K\left(\frac{q\left(Q + q\frac{a}{y}\right)}{y^2} - q\frac{\left(q\frac{a}{y}\right)}{\left(y - \frac{a^2}{y}\right)^2}\right)\frac{\vec{y}}{y}$$

$$\vec{F}_q = \frac{Kq}{y^2}\left(\left(Q + q\frac{a}{y}\right) - \frac{\left(q\frac{a}{y}\right)}{\left(1 - \frac{a^2}{y^2}\right)^2}\right)\frac{\vec{y}}{y} \tag{3.23}$$

which can be simplified as follows:

$$\vec{F}_q = \frac{Kq}{y^2}\left(Q - \frac{q\left(\frac{a}{y}\right)^3\left(2 - \frac{a^2}{y^2}\right)}{\left(1 - \frac{a^2}{y^2}\right)^2}\right)\frac{\vec{y}}{y} \tag{3.24}$$

We will discuss some of the special cases as follows:

(a) **Long Distance**: $y \gg a$.

The expression for force reduces as

$$\vec{F}_q = \frac{Kq}{y^2} Q \frac{\vec{y}}{y} \tag{3.25}$$

This is similar to the conventional Coulomb's law applied to two small charged particles. However, the force is altered due to the influence of the induced charge distribution present on the surface of a sphere.

(b) **Short Distance**: $y \approx a$, then the Eq. (3.24) reduces to

$$\vec{F}_q = \frac{Kq}{y^2} \left(Q - q - 2q'\frac{a}{y} \right) \tag{3.26}$$

However, $-q - 2q'\frac{a}{y} \gg Q$. As a result, $F_q \propto \frac{1}{y^2}$ and it is important to note here that the force becomes attractive at very close distances.

This problem explains a general property that why an excess surface charge does not immediately move on account of mutual repulsion of individual charges. The moment the element of surface charge is removed, the image charge pulls it back. However, it requires an ample work that is to be done in order to move the surface charge to infinity.

3.4 Influence of a Uniform Electric Field on Conducting Sphere

Let us envisage a conducting sphere with a radius 'a' immersed in a uniform electric field. This uniform field can be thought of as originating from a pair of charges—one positive and one negative positioned infinitely far away. To facilitate our understanding, let us place charges of magnitude $+Q$ at $Z = -R$ and $-Q$ at $Z = +R$, as illustrated in the accompanying diagram. This configuration allows us to analyze the influence of these charges on the electric field experienced at a specific point, denoted as o. The resulting electric field at point o can be described mathematically as follows:

$$E = \frac{KQ}{R^2} + \frac{KQ}{R^2} = 2\frac{KQ}{R^2} \tag{3.27}$$

As $E \to E_0$; $Q \to \infty$, $R \to \infty$ implies that $E_0 \to 2\frac{KQ}{R^2}$.

We now assume that the conducting sphere is placed at the origin and, therefore, the potential will be due to the charges $\pm Q$ located at positions $Z = \mp R$ and their images $q_1 = \frac{-Qa}{R}$ at $z = \frac{-a^2}{R}$ and $q_1 = \frac{Qa}{R}$ at $Z = \frac{+a^2}{R}$ and OP $= r$ (Fig. 3.5).

Therefore, the net potential is due to charges $Q, -Q, q_1$, and q_2

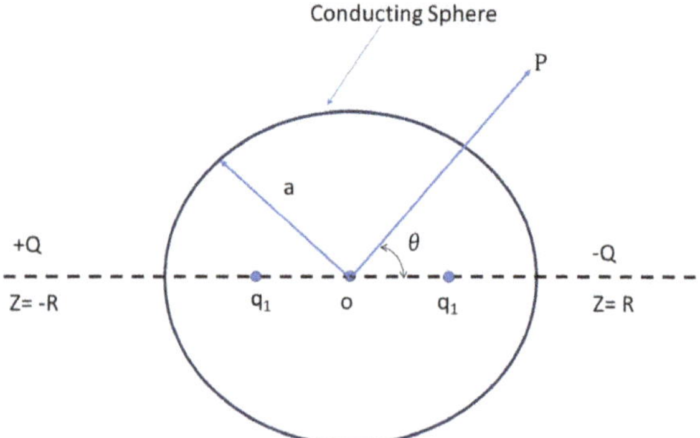

Fig. 3.5 Analysis of a conducting sphere placed in a uniform electric field using the method of image charges

$$V(\vec{r}) = K\left(\frac{Q}{|\vec{r} - \vec{r}_Q|} + \frac{q}{|\vec{r} - \vec{r}_{q_1}|} + \frac{q}{|\vec{r} - \vec{r}_{q_2}|} + \frac{-Q}{|\vec{r} - \vec{r}_Q|}\right) \tag{3.28}$$

Or we can write it as follows:

$$\frac{V(\vec{r})}{K} = \frac{Q}{|\vec{r} + \vec{R}|} - \frac{Q}{|\vec{r} - \vec{R}|} + \frac{\frac{-Qa}{R}}{|\vec{r} + \frac{a^2}{R}|} + \frac{\frac{Qa}{R}}{|\vec{r} - \frac{a^2}{R}|}$$

$$\frac{V(\vec{r})}{K} = \frac{Q}{\left(r^2 + R^2 + 2rR\cos(\theta)\right)^{1/2}} + \frac{-Q}{\left(r^2 + R^2 - 2rR\cos(\theta)\right)^{1/2}}$$

$$+ \frac{-Q\frac{a}{R}}{\left(r^2 + \frac{a^4}{R^2} + 2r\frac{a^2}{R}\cos(\theta)\right)^{1/2}} + \frac{Q\frac{a}{R}}{\left(r^2 + \frac{a^4}{R^2} - 2r\frac{a^2}{R}\cos(\theta)\right)^{1/2}} \tag{3.29}$$

The variable V has been represented using the spherical coordinates of the observation point. Additionally, in the first two terms of the equation mentioned above, the condition $r \ll R$ holds true. We, therefore, factor out R^2 and expand the radicals by Binomial Theorem. In a similar manner, for the third and fourth terms, we can factor out r^2 and upon expanding, we obtain

$$\left(r^2 + R^2 + 2rR\cos(\theta)\right)^{-1/2} = R^{-1}\left(1 + \frac{r^2}{R^2} + \frac{2r}{R}\cos(\theta)\right)^{-1/2} \tag{3.30}$$

$$\left(r^2 + R^2 + 2rR\cos(\theta)\right)^{-1/2} = R^{-1}\left(1 - \frac{1}{2}\frac{r^2}{R^2} - \frac{r}{R}\cos(\theta) - \cdots\right)$$

and

$$\left(r^2 + \frac{a^4}{R^2} + 2r\frac{a^2}{R}\cos(\theta)\right)^{-1/2} = r^{-1}\left(1 + \frac{a^4}{R^2r^2} + 2\frac{a^2}{rR}\cos(\theta)\right)^{-1/2}$$

$$= r^{-1}\left(1 - \frac{a^4}{2R^2r^2} - \frac{a^2}{rR}\cos(\theta) - \cdots\right) \tag{3.31}$$

(since $\frac{a}{r}$ and $\frac{a}{R}$ are less than 1). Therefore, on substuting Eqs. (3.30) and (3.31) in Eq. (3.29), we get

$$\frac{V(\vec{r})}{K} = \frac{Q}{R}\left(1 - \frac{r^2}{2R^2} - \frac{r}{R}\cos(\theta) + \cdots\right) - \frac{Q}{R}\left(1 - \frac{r^2}{2R^2} + \frac{r}{R}\cos(\theta) + \cdots\right)$$

$$- \frac{Qa}{Rr}\left(1 - \frac{a^4}{2R^2r^2} - \frac{a^2}{rR}\cos(\theta) + \cdots\right) + \frac{Qa}{Rr}\left(1 - \frac{a^4}{2R^2r^2} + \frac{a^2}{rR}\cos(\theta) + \cdots\right)$$

$$= \frac{Q}{R}\left(-2\frac{r}{R}\cos(\theta) + \cdots\right) + \frac{Qa}{Rr}\left(2\frac{a^2}{Rr}\cos(\theta) + \cdots\right) \tag{3.32}$$

as $Q \to \infty$ and $R \to \infty$ implies $\frac{Q}{R^2} \to 0$. Under these approximations the above expression for potential assumes the following form:

$$V(\vec{r}) = \frac{-2KQ}{R^2}\left(r - \frac{a^3}{r^2}\right)\cos(\theta) \tag{3.33}$$

In the aforementioned equation, the component $-\vec{E}_0.\vec{r}$ signifies the potential generated by a uniform electric field. Conversely, the term $E_0\frac{a^3}{r^2}\cos(\theta)$ corresponds to the potential created by the induced surface charge density, which can also be understood in terms of the influence of image charges.

Induced Charge: The induced surface charge density is calculated as follows:

$$\sigma = \frac{-1}{4\pi K}\frac{\partial V}{\partial r}\Big|_{r=a}$$

$$\sigma = -\frac{1}{4\pi K}\left(-E_0\left(1 + \frac{2a^3}{r^3}\right)\cos(\theta)\right)_{r=a}$$

$$\sigma = \frac{E_0}{4\pi K}(1 + 2)\cos(\theta) = \frac{3E_0}{4\pi K}\cos(\theta) \tag{3.34}$$

It is important to mention here that the surface integral of this expression vanishes and hence there is no difference between grounded and an insulated sphere.

Example 3.1 A conducting sphere of radius R is placed in a uniform electric field $\vec{E}_0$ directed along $+ z$-axis. The electric potential for outside points is given as $V = -E_0\left(1 - \frac{R^3}{r^3}\right)r\cos\theta$, where r is the distance from the centre and θ is the polar angle. Calculate the charge density on the surface of the sphere.

Solution:

Since,

$$V = -E_0 \left(1 - \frac{R^3}{r^3} \right) r \cos\theta$$

The charge density is related to $\vec{E}_0$ by a relation

$$E = \frac{\sigma}{\epsilon_0}$$

Further,

$$\vec{E} = -\vec{\nabla}V$$

$$\sigma = -\epsilon_0 \frac{\partial v}{\partial r}\Big|_{r=R} = 3\epsilon_0 E_0 \cos\theta$$

Example 3.2 An electric dipole with dipole moment $\vec{p}$ is located at a separation d above an infinite grounded conducting plane. Determine the potential and force on the dipole.

Solution:

The image dipole $p' = p$ is located at $z = -d$.

The potential due to both dipoles is $V(z) = \frac{1}{4\pi\epsilon_0} \left(\frac{\vec{p}\cdot\hat{z}}{(z-d)^2} + \frac{-\vec{p}\cdot\hat{z}}{(z+d)^2} \right)$

$$V(z) = \frac{\vec{p}\cdot\hat{z}}{4\pi\epsilon_0} \left(\frac{1}{(z-d)^2} - \frac{1}{(z+d)^2} \right)$$

The force on the dipole is $\vec{F} = -\vec{\nabla}V$

$$\frac{\partial V}{\partial z} = \frac{\vec{p}\cdot\hat{z}}{4\pi\epsilon_0} \left(-\frac{2}{(z-d)^3} + \frac{2}{(z+d)^3} \right)$$

$$F = 2\left(\frac{\vec{p}\cdot\hat{z}}{4\pi\epsilon_0} \left(\frac{1}{(z-d)^3} - \frac{1}{(z+d)^3} \right) \right)$$

Example 3.3 The region between two concentric right circular cylinders contains a uniform charge density ρ. Calculate the potentialis V.

Solution:

Using the Poisson's equation

$$\frac{1}{r}\frac{d}{dr}\left(r\frac{dV}{dr}\right) = -\frac{\rho}{\epsilon}$$

$$\frac{d}{dr}\left(r\frac{dV}{dr}\right) = -\frac{\rho r}{\epsilon}$$

Integrating, we get

$$r\frac{dV}{dr} = -\frac{\rho r^2}{2\epsilon} + A$$

$$\frac{dV}{dr} = -\frac{\rho r}{2\epsilon} + \frac{A}{r}$$

Again integrating, we obtain

$$V = \frac{\rho r^2}{4\epsilon} + A ln(r) + B$$

Example 3.4 Two equal point charges $+q$ and $-q$ are placed at a distance d apart, at a distance h from an infinite grounded conducting wall. Find the potential and force.

Solution:

We can place $-q$ at $z = -h$.

The potential is $V(r) = \frac{q}{4\pi\epsilon_0 r_+} - \frac{q}{4\pi\epsilon_0 r_-}$.

where $r_+ = \sqrt{(x-x_1)^2 + (y-y_1)^2 + (z-h)^2}$.

and $r_- = \sqrt{(x-x_1)^2 + (y-y_1)^2 + (z+h)^2}$.

The force on charge $+q$ is

$$\vec{F} = -\vec{\nabla}V$$

$$F_x = -\frac{\partial V}{\partial x},$$

$$F_y = -\frac{\partial V}{\partial y}; \quad F_z = -\frac{\partial V}{\partial z}$$

Example 3.5 Find the energy stored in a uniformly charged solid sphere of radius R and charge.

Solution:

$\vec{E} = \frac{1}{4\pi\epsilon_0 R^3}\frac{qr}{R^3}\hat{r}, r > R$ and $\vec{E} = \frac{1}{4\pi\epsilon_0 r^2}\frac{q}{r^2}\hat{r}, r < R$

$$W = \frac{\epsilon^2}{2} \frac{q^2}{(4\pi\epsilon_0)^2} \left\{ \int_R^\infty \frac{1}{r^4}(r^2 4\pi\,dr) + \int_0^R \left(\frac{r}{R^3}\right)^2 (r^2 4\pi\,dr) \right\}$$

$$W = \frac{1}{4\pi\epsilon_0} \frac{3q^2}{5R}$$

Example 3.6 A charge $q = 4\,\mu C$ is positioned at 8 cm away from the centre of a conducting grounded sphere of radius $a = 3\,cm$. Calculate the electrostatic force between the charge and the image charge.

Solution:

We know that image charge is given by

$$q' = -q\frac{a}{y} = -4 \times 10^{-6}\frac{3}{8} = -1.5\,\mu C$$

The position y' of charge q' is

$$y' = \frac{a^2}{y} = \frac{3^2}{8} = 1.125\,cm$$

The distance between q and q' is given by

$$r = y - y' = 8 - 1.125 = 6.875\,cm = 0.06875\,m.$$

The force between q and q' is

$$F = \frac{Kq_1 q_2}{r^2} = \frac{9 \times 10^9 \times 4 \times 10^{-6} \times 1.5 \times 10^{-6}}{(0.06875)^2} = 11.42\,N$$

Example 3.7 Calculate the work required to bring a $3\,\mu C$ point charge from infinity to a point 7 cm from the centre of conducting grounded sphere of radius 3 cm.

Solution:

The image charge $q' = -q\frac{a}{y} = -3 \times 10^{-6}\frac{3}{7} = -1.29\,\mu C.$

Image position $y' = \frac{a^2}{y} = \frac{3^2}{7} = 1.29\,cm.$
The potential energy at a distance $y = 7\,cm$

$$W = \frac{Kqq'}{y - y'} \approx 0.61\,J$$

Example 3.8 A point charge $q = 2\,\mu C$ is placed 8 cm away from the centre of an insulated conducting sphere with radius $a = 3\,cm$ and total charge $Q = 5\,\mu C$.

Calculate the electric potential at a point P located 12 cm from the centre of the sphere, along the line joining the charge q and the centre of the sphere.

Solution:

Image charge $q' = -q\frac{a}{y} = -2 \times 10^{-6}\frac{3}{8} = -0.75\,\mu C$.

Image position $y' = \frac{a^2}{y} = 1.125\,cm$.

The effective charge q'' on the sphere due to the presence of Q and q'

$$q'' = Q + q\frac{a}{y} = 5.75\,\mu C$$

The total potential at point P due to q, q' and q'' is

$$V(P) = 9 \times 10^9 \left(\frac{2 \times 10^{-6}}{0.4} + \frac{-0.75 \times 10^{-6}}{0.10875} + \frac{5.75 \times 10^{-6}}{0.12} \right) = 819180\,V$$

Example 3.9 A conducting sphere of radius $a = 4\,cm$ is placed in uniform electric field $\vec{E}_0 = 3 \times 10^5\,V/m$. Calculate the induced surface charge density σ at points on the sphere's surfaces at angles $\theta = 0^0$, $\theta = 45^0$ and $\theta = 90^0$.

Solution:

The induced surface charge density on a conducting sphere in a uniform electric field is given by

$$\sigma = \frac{3E_0}{4\pi K}Cos\theta$$

For different θ, σ is given by
For $\theta = 0^0$

$$\sigma = \frac{3 \times 3 \times 10^5}{4\pi \times 9 \times 10^9} \times 1 = 7.96 \times 10^{-6}\,C/m^2$$

For $\theta = 45^0$

$$\sigma = \frac{3 \times 3 \times 10^5}{4\pi \times 9 \times 10^9} \times Cos45^0 = 5.63 \times 10^{-6}\,C/m^2$$

For $\theta = 90^0$

$$\sigma = \frac{3 \times 3 \times 10^5}{4\pi \times 9 \times 10^9} \times Cos90^0 = 0\,C/m^2$$

Example 3.10 Given a conducting sphere of radius $a = 5$ cm in a uniform electric field $E_0 = 2 \times 10^5$ V/m. Calculate the potential at a point P located $r = 15$ cm from the centre of sphere along the direction of electric field.

Solution:

The potential outside the sphere at a distance r along the direction of the electric field is given by

$$V(r) = -E_0 r \cos\theta + E_0 \frac{a^3}{r^2} \cos\theta$$

Here, $\theta = 0^0$, $E_0 = 2 \times 10^5$ V/m, $r = 15$ cm $= 0.15$ m and $a = 5cm = 0.05m$

$$V(r) = -28888.89 \text{ V}$$

3.5   Potential Due to Dipole

It is observed that potential becomes zero either when $r \to \infty$ or $Q \to 0$. However, potential due to dipole is non-zero even if $Q = 0$. Consider a dipole as show below (Fig. 3.6).

Since potential obeys superposition principle, therefore, potential at point P is given by

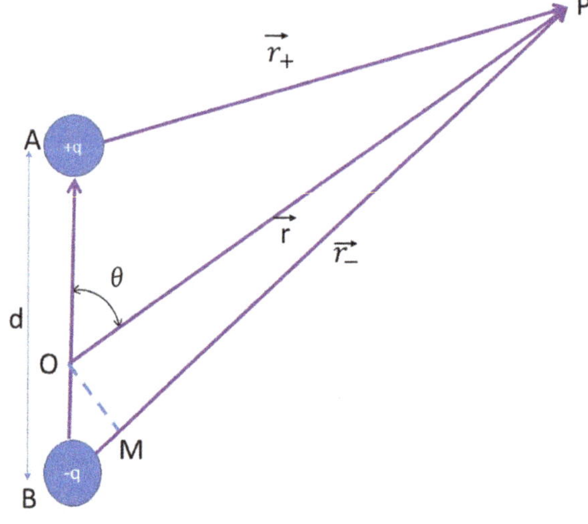

Fig. 3.6 Schematic representation of an electric dipole

$$V(\vec{r}) = \frac{1}{4\pi\epsilon_0}\left(\frac{q}{r_+} - \frac{q}{r_-}\right) \tag{3.35}$$

From the law of cosine, we can write

$$r_{\pm}^2 = r^2 + \left(\frac{d}{2}\right)^2 \mp rd\cos(\theta)$$

$$r_{\pm}^2 = r^2\left(1 \mp \frac{d}{r}\cos(\theta) + \frac{d^2}{4r^2}\right)$$

In the regime $r \gg d$, i.e., the observation point is far off from the dipole

$$r_{\pm} \cong r\left(1 \mp \frac{d}{r}\cos(\theta)\right)^{1/2}$$

$$\frac{1}{r_{\pm}} \cong \frac{1}{r}\left(1 \mp \frac{d}{r}\cos(\theta)\right)^{-1/2}$$

$$\frac{1}{r_{\pm}} \cong \frac{1}{r}\left(1 \pm \frac{d}{2r}\cos(\theta)\right)$$

Thus,

$$\frac{1}{r_+} - \frac{1}{r_-} \cong \frac{d}{r^2}\cos(\theta) \tag{3.36}$$

Hence, electric potential due to an electric dipole is written as follows:

$$V(\vec{r}) \cong \frac{1}{4\pi\epsilon_0}\frac{qd}{r^2}\cos(\theta) \tag{3.37}$$

We know that potential due to the point charge $V(\vec{r}) \propto \frac{1}{r}$, here we found that potential due to the dipole as $V(\vec{r}) \propto \frac{1}{r^2}$. This motivates us to expand $\frac{1}{r}$ by binominal expansion.

3.6 Expansion of Multipole Moments

Let us examine the charge distribution depicted in the illustration. Our goal is to determine the electric potential at point P (Fig. 3.7).

Therefore, the potential at point P can be written explicitly as follows:

$$V(\vec{r}) = \frac{1}{4\pi\epsilon_0}\int \frac{1}{R}\rho(\vec{r}')d\tau'$$

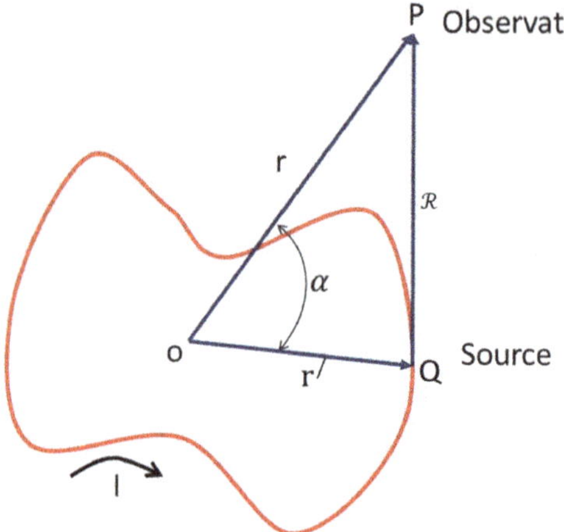

Fig. 3.7 Diagrammatic representation of certain charge distribution in a region

$$R^2 = r^2 + r'^2 - 2rr'\cos(\alpha)$$

$$R^2 = r^2\left(1 + \left(\frac{r'}{r}\right)\left(\frac{r'}{r} - 2\cos(\alpha)\right)\right) \tag{3.38}$$

We can write the expression as

$$R = r\sqrt{1 + \epsilon} \tag{3.39}$$

where

$$\epsilon = \left(\frac{r'}{r}\right)\left(\frac{r'}{r} - 2\cos(\alpha)\right) \tag{3.40}$$

If $\epsilon < 1$, we get from Eq. (3.39) after expansion

$$\frac{1}{R} = \frac{1}{r}(1 + \epsilon)^{-1/2}$$

$$\frac{1}{R} = \frac{1}{r}\left(1 - \frac{1}{2}\epsilon + \frac{3}{8}\epsilon^2 - \frac{5}{16}\epsilon^3 + \cdots\right)$$

Substitute value of ϵ in the equation, we obtain

$$\frac{1}{R} = \frac{1}{r}\left(1 - \frac{1}{2}\left(\frac{r'}{r}\right)\left(\frac{r'}{r} - 2\cos(\alpha)\right) + \frac{3}{8}\left(\frac{r'}{r}\right)^2\left(\frac{r'}{r} - 2\cos(\alpha)\right)^2\right.$$

$$\left. -\frac{5}{16}\left(\frac{r'}{r}\right)^3\left(\frac{r'}{r} - 2\cos(\alpha)\right)^3 + \cdots\right)$$

$$\frac{1}{R} = \frac{1}{r}\left[1 - \frac{1}{2}\left(\frac{r'}{r}\right)^2 + \left(\frac{r'}{r}\right)\cos(\alpha) + \frac{3}{8}\left(\frac{r'}{r}\right)^2 \right.$$

$$\left(\left(\frac{r'}{r}\right)^2 + 4\cos^2(\alpha) - \left(\frac{r'}{r}\right)4\cos(\alpha) \right) - \frac{5}{16}\left(\frac{r'}{r}\right)^3$$

$$\left. \left(\left(\frac{r'}{r}\right)^3 - 6\left(\frac{r'}{r}\right)^2\cos(\alpha) + 12\left(\frac{r'}{r}\right)\cos^2(\alpha) - 8\cos^3(\alpha) \right) + \cdots \right]$$

$$\frac{1}{R} = \frac{1}{r}\left[1 + \left(\frac{r'}{r}\right)\cos(\alpha) + \left(\frac{r'}{r}\right)^2\frac{(3\cos^2(\alpha) - 1)}{2} \right.$$

$$\left. + \left(\frac{r'}{r}\right)^3\frac{(5\cos^2(\alpha) - \cos(\alpha))}{2} + \cdots \right]$$

Coefficients of $\left(\frac{r'}{r}\right)$ are Legendre polynomials

$$\frac{1}{R} = \frac{1}{r}\sum_{n=0}^{\infty}\left(\frac{r'}{r}\right)^n P_n(\cos(\alpha)) \tag{3.41}$$

We know that $P_0(\cos(\alpha)) = 1; P_1(\cos(\alpha)) = \cos(\alpha)$.
Hence, expression for potential becomes

$$V(\vec{r}) = \frac{1}{4\pi\epsilon_0}\sum_{n=0}^{\infty}\frac{1}{r^{n+1}}\int (r')^n P_n\cos(\alpha)\rho(\vec{r}')d\tau' \tag{3.42}$$

Figure 3.8 likely illustrates the geometric relationship between the vectors $\vec{r}, \vec{r}'$, and the angle α. It may also show how the angle α influences the contributions of the monopole, dipole, quadrupole and higher-order terms to the potential $V(\vec{r})$.

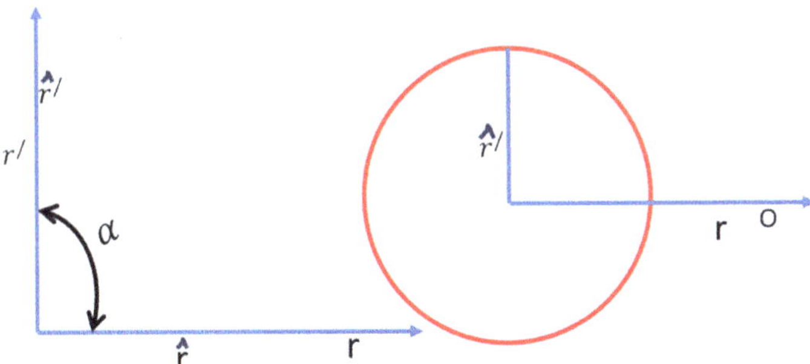

Fig. 3.8 Representation of radial vectors r and $r/$

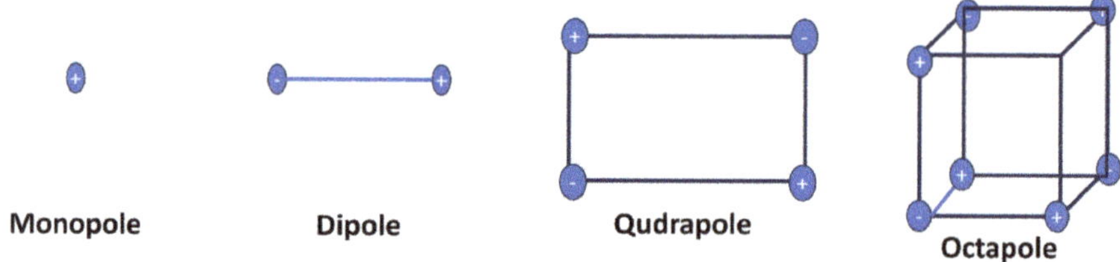

Fig. 3.9 Charge configurations representing monopole, dipole, quadrupole, and octopole arrangements, showing increasing spatial complexity and multipole order

This expression can be explicitly written as under

$$V(\vec{r}) = \frac{1}{4\pi \epsilon_0}\left[\frac{1}{r}\int \rho(\vec{r}')d\tau' + \frac{1}{r^2}\int r'\cos(\alpha)\rho(\vec{r}')d\tau'\right.$$
$$\left. + \frac{1}{r^3}\int (r')^2\frac{(3\cos^2(\alpha)-1)}{2}\rho(\vec{r}')d\tau' + \cdots\right]$$

which represents the multipole expansion. The first term on R.H.S represents monopole term, the second term represents the dipole term and so on (Fig. 3.9)

$$V(\vec{r}) = \text{Monopole} + \text{Dipole} + \text{Qudrupole} + \text{Octapole} + \cdots$$

The Monopole and Dipole Terms: Dominating term at large $\vec{r}$

$$V_{\text{Mon.}}(\vec{r}) = \frac{1}{4\pi \epsilon_0}\frac{Q}{r} \tag{3.43}$$

where

$$Q = \int \rho(\vec{r}')d\tau'$$

$V_{\text{Mon.}}(\vec{r})$ is the exact potential for point charge. However, if the total charge is zero the dominant term is

$$V_{\text{Dip.}}(\vec{r}) = \frac{1}{4\pi \epsilon_0}\frac{1}{r^2}\int r'\cos(\alpha)\rho(\vec{r}')d\tau' \tag{3.44}$$

So, we can write $r'\cos(\alpha) = \hat{r}\cdot\vec{r}'$. Therefore $V_{\text{Dip.}}(\vec{r})$ can be written as:

$$V_{\text{Dip.}}(\vec{r}) = \frac{1}{4\pi \epsilon_0}\frac{1}{r^2}\hat{r}\cdot\int \vec{r}'\rho(\vec{r}')d\tau' \tag{3.45}$$

Here, we define dipole moment

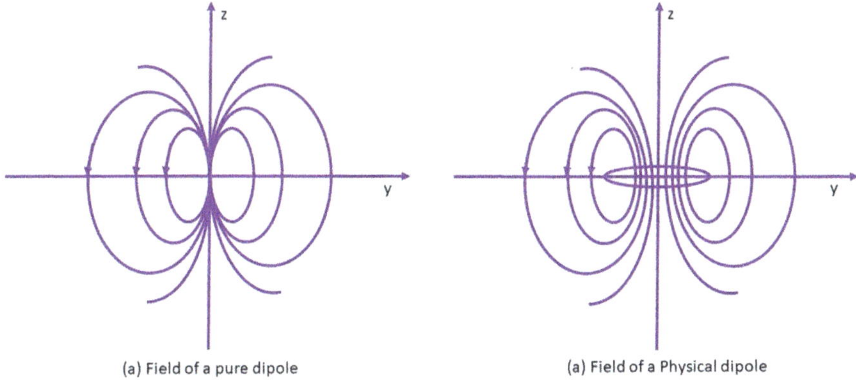

Fig. 3.10 Comparison of electric field lines of a pure dipole (idealized point dipole) and a physical dipole (finite separation between charges). The pure dipole field (left) exhibits perfect radial symmetry, while the physical dipole field (right) reflects the finite spatial extent of the charge separation

$$\vec{P} = \int \vec{r}\,'\rho(\vec{r}\,')d\tau'$$
(3.46)

This is the dipole moment of the exact dipole or ideal dipole ($d \to 0$).
Therefore, we can write

$$V_{\text{Dip.}}(\vec{r}) = \frac{1}{4\pi\epsilon_0}\frac{\vec{P}\cdot\hat{r}}{r^2}$$
(3.47)

The dipole moment is governed by the geometry of charge distribution for collection of charges $\vec{P} = \sum_{i=1}^{n} q_i\vec{r}_i'$ (Fig. 3.10).

For Physical Dipole: For a physical dipole consisting of two equal and opposite charges separated by a small distance, the dipole moment is defined as

$$\vec{P} = q\vec{r}_+' - q\vec{r}_-'$$

For small separations, this simplifies to

$$\vec{P} = q\vec{d}$$
(3.48)

Where $\vec{d}$ is the displacement vector from the negative charge to the positive charge. This Eq.(3.48) represents the fundamental definition of a physical dipole moment.

3.7 Expanding the Vector Potential Using Multipole Moments

We can derive an approximate expression for the vector potential of a localized charge distribution at any given point by utilizing a multipole expansion technique. The idea here is to make a power series expansion in $\frac{1}{r}$, where r is the distance to the observation point.

When, r is large enough, the series will primarily be influenced by the smallest non-zero terms and the higher terms can be ignored.

$$\frac{1}{R} = \frac{1}{\sqrt{r^2 + r'^2 - 2rr'\cos(\alpha)}} = \frac{1}{r}\sum_{n=0}^{\infty}\left(\frac{r'}{r}\right)^n P_n(\cos(\alpha)) \qquad (3.49)$$

where α is the angle between $\vec{r}$ and $\vec{r}'$. Accordingly, the vector potential of a current loop can be written:

$$\vec{A}(\vec{r}) = \frac{\mu_0 I}{4\pi}\int\frac{1}{R}\overrightarrow{dl} = \frac{\mu_0 I}{4\pi}\sum_{n=0}^{\infty}\frac{1}{r^{n+1}}\int(r')^n P_n\cos(\alpha)\rho(\vec{r}')\overrightarrow{dl}' \qquad (3.50)$$

More explicitly we can write above equation as follows (Fig. 3.11):

$$\vec{A}(\vec{r}) = \frac{\mu_0 I}{4\pi}\left[\frac{1}{r}\oint\overrightarrow{dl}' + \frac{1}{r^2}\oint r'\cos(\alpha)\overrightarrow{dl}' + \frac{1}{r^3}\oint (r')^2\left(\frac{3}{2}\cos^2\alpha - \frac{1}{2}\right)\overrightarrow{dl}' + \cdots\right]$$
$$(3.51)$$

When r is extremely large, the expansion will primarily be influenced by the smallest non-zero contributions. From the notion of multipole expansion of potential V, we deduce that the first term behaves like $\frac{1}{r}$ (the monopole term), the second term behaves like $\frac{1}{r^2}$ (the dipole term), the third term behaves like $\frac{1}{r^3}$ (the quadrupole term) and so on and so forth. Notably, the magnetic term is always zero, as the integral represents the total vector displacement around a closed loop, specifically $\oint\overrightarrow{dl}' = 0$.

Fig. 3.11 Schematic representation for multipole expansion of moments

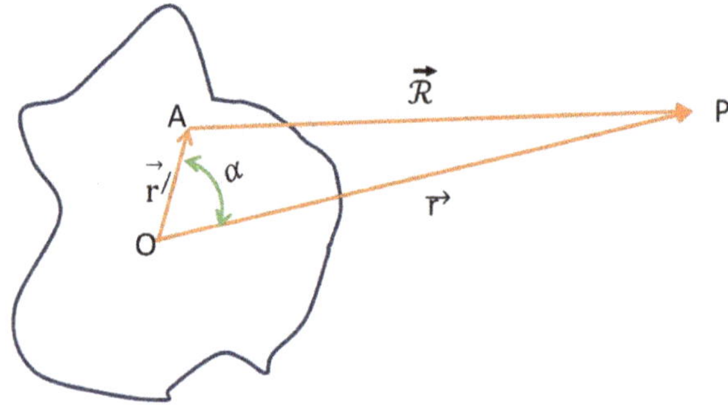

This indicates the absence of magnetic monopoles in nature, which aligns with the assumption in Maxwell's equations, where $\vec{\nabla} \cdot \vec{B} = 0$. Consequently, the dominant term in this context is the dipole term.

$$\vec{A}(\vec{r})_{\text{Dipole}} = \frac{\mu_0 I}{4\pi} \frac{1}{r^2} \oint r' \cos(\alpha) \vec{dl'} = \frac{\mu_0 I}{4\pi r^2} \oint \hat{r} \cdot \vec{r'} \vec{dl'} \tag{3.52}$$

The integral in the above equation can be written as:

$$\oint \hat{r} \cdot \vec{r'} \vec{dl'} = -\hat{r} \times \int d\vec{a}$$

$$\vec{A}(\vec{r})_{\text{Dipole}} = \frac{\mu_0 I}{4\pi r^2} \int d\vec{a} \times \hat{r}$$

$$\vec{A}(\vec{r})_{\text{Dipole}} = \frac{\mu_0}{4\pi} \frac{\vec{m} \times \hat{r}}{r^2} \tag{3.53}$$

The magnetic moment, denoted $\vec{m}$, is given by $\vec{m} = I \int d\vec{a} = I\vec{a}$ where I is the current, and $\vec{a}$ represents the vector area of the loop through which the current flows. When the loop lies flat, $\vec{a}$ simplifies to the standard scalar area enclosed by the loop, with its direction determined by the right-hand rule pointing along the axis that aligns with the current's circulation.

From Eq. (3.53), we observe that the magnetic dipole moment is unaffected by the location of the origin. This contrasts with the electric dipole moment, which is origin-independent only in cases where the system has a net-zero charge. Given that magnetic monopoles are not observed (i.e., the magnetic monopole moment is zero), it follows naturally that the magnetic dipole moment is consistently independent of the origin. Additionally, the magnetic dipole term tends to be the leading term in a multipole expansion (unless $m = 0$), making it a reliable approximation for the true potential field in most practical cases.

The magnetic field of an ideal dipole configuration is most easily calculated by placing $\vec{m}$ at the origin and orienting it along the z-axis. Under these conditions, the potential at a point (r, θ, ϕ) in spherical coordinates can be derived for a straightforward analysis.

In line with Maxwell's equation, $\vec{\nabla}.\vec{B} = 0$ which is a foundational principle of vector potential theory, the magnetic dipole term becomes the primary component in field approximations, as higher-order terms contribute minimally under typical circumstances (Fig. 3.12).

The vector potential at point (r, θ, ϕ) is

$$\vec{A}(\vec{r})_{\text{Dipole}} = \frac{\mu_0}{4\pi} \frac{m \sin(\theta)}{r^2} \hat{\phi} \tag{3.54}$$

Hence, the magnetic field can be written as

$$\vec{B} = \vec{\nabla} \times \vec{A}$$

Fig. 3.12 A magnetic dipole aligned along Z-axis

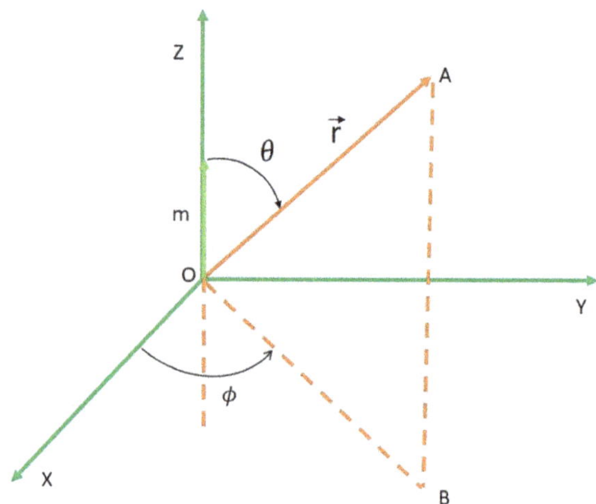

$$\vec{B} = \frac{\mu_0 m}{4\pi}\left(2\cos(\theta)\hat{r} + \sin(\theta)\hat{\theta}\right) \tag{3.55}$$

This equation is reminiscent to the field of an electric dipole.

Example 3.11 A conducting sphere of radius $a = 2$cm is placed in a uniform electric field $E_0 = 1.5 \times 10^5 \text{V/m}$. Determine the force on the induced dipole moment of the sphere due to the external field.

Solution:

The induced dipole moment p of a conducting sphere in an external field E_0 is given as follows:

$$p = 4\pi\epsilon_0 a^3 E_0$$

We know that

$$\epsilon_0 = 8.85 \times 10^{12}\text{C}^2/\text{Nm}^2$$
$$a = 2\,\text{cm} = 0.02\,\text{m}$$
$$E_0 = 1.5 \times 10^5\,\text{V/m}$$
$$p \approx 1.33 \times 10^{-10}\,\text{C}^2V/N$$

The force F on the induced dipole moment p in the uniform electric field E_0 is given by

$$F = \vec{P} \cdot \vec{\nabla}E_0$$

Since, E_0 is uniform hence the force $F = 0$.

Example 3.12 A conducting sphere of radius $a = 3$ cm is placed in a uniform electric field $E_0 = 4 \times 10^5$ V/m. Calculate the work done to remove the sphere from the electric field.

Solution:

The dipole moment p, induced by the sphere in presence of electric field E_0 is

$$p = 4\pi \epsilon_0 a^3 E_0$$
$$\epsilon_0 = 8.85 \times 10^{12} C^2/Nm^2$$
$$a = 3\,\text{cm} = 0.03\,\text{m}$$
$$E_0 = 4 \times 10^5\,\text{V/m}$$
$$p \approx 12.0 \times 10^{-10}\,C^2 V/N$$

The potential energy U of the dipole in the electric field E_0 is

$$U = -pE_0 = -4.8 \times 10^{-4}\text{J}$$

The necessary work done to remove the sphere from the electric field is $W = -U = 4.8 \times 10^{-4}\text{J}$.

Example 3.13 A dipole with charges $\pm q = 1 \times 10^6$ C separated by a distance $d = 2 \times 10^{-2}$ m is oriented along z-axis. Determine the electric potential V at points located at.

$$r = 0.1\,\text{m},\ \theta = 0^0$$
$$r = 0.1\,\text{m},\ \theta = 90^0$$

Solution:

The potential due to a dipole at a distance r and angle θ from the axis is given by

$$V(r, \theta) = \frac{1}{4\pi \epsilon_0} \frac{qd \cos \theta}{r^2}$$

For $\theta = 0^0$ (along the dipole axis)

$$V(0.1, 0^0) = 1.8 \times 10^{16}\,\text{V}$$

For $\theta = 90^0$ (perpendicular to the dipole axis)

$$V(0.1, 90^0) = 0$$

Example 3.14 Consider a quadrupole with charges $+q, -q, +q, -q$ located at positions along the z-axis, $\pm d$ and $\pm 2d$ where $q = 2 \times 10^{-6}$ C and $d = 5 \times 10^{-2}$ m.

Deduce the approximate potential at a point $r = 1$ m far away from the quadrupole, along the z-axis.

Solution:

For a quadrupole, the leading term in the potential at large distance $r \gg d$ can be given by

$$V(r, \theta) = \frac{1}{4\pi \epsilon_0} \frac{Q}{r^3} P_2(\cos \theta)$$

where $Q = \sum qz^2$ for the quadrupole moment and $P_2(\cos \theta) = \frac{3\cos^2 \theta - 1}{2}$
The charges are positoned along the z-axis as follows:

- $+q$ at $z = +d$, $-q$ at $z = +2d$, $+q$ at $z = -d$ and $-q$ at $z = -2d$

$$Q = -6qd^2$$

Given that, $q = 2 \times 10^{-6}$ C and $d = 5 \times 10^{-2}$ m.

$$Q = -3.0 \times 10^{-8} \text{ Cm}^2.$$
$$r = 1 \text{ m and } \theta = 0^0$$
$$V(1, 0^0) = -270 \text{ V}$$

Example 3.15 A circular loop of radius $a = 0.1$ m possesses a current $I = 5$ A. Determine the approximate vector potential $\vec{A}$ at a point located at a distance $r = 1$ m from the loop along its axis using the dipole term in the multipole expansion.

Solution:

We know that

$$\vec{A}_{\text{Dipole}} = \frac{\mu_0 I}{4\pi r^2} \int r' \cos \alpha dl'$$

For a circular loop dipole moment $m = I\pi a^2 = 5 \times 3.14 \times (0.1)^2 = 0.157 \text{ Am}^2$
Since $\hat{r}$ is along z-axis, $\vec{m} \times \hat{r}$ points in the azimuthal direction

$$\vec{A}_{\text{Dipole}} = \frac{4\pi \times 10^{-7} \times 0.157}{4\pi \times (1)^2} = 1.57 \times 10^{-8} \text{ Tm}$$

Example 3.16 Using the magnetic dipole moment $m = 0.157 \text{ Am}^2$ from an Example 3.15, calculate the magnetic field $\vec{B}$ at a point $r = 1$ m along the direction of dipole $(\theta = 0^0)$.

Solution:

The magnetic field $\vec{B}$ in terms of the dipole moment is given by

$$\vec{B} = \frac{\mu_0 m}{4\pi r^3}\left(2\cos\theta\,\hat{r} + \sin\theta\,\hat{\theta}\right)$$

Along the axis $\theta = 0^0$ $\vec{B}$ is given as

$$B_r = \frac{\mu_0 m}{4\pi r^3}\left(2\cos 0^0\right) = 3.14 \times 10^{-8}T$$

Example 3.17 For a circular current loop of radius $a = 0.2$ m with current $I = 10$ A, deduce the vector potential $\vec{A}$ approximately at a point $r = 1$ m located at an angle $\theta = 45^0$ from the loop axis using the dipole term.

Solution:

Magnetic dipole moment is

$$m = I\pi a^2 = 10 \times 3.14 \times (0.2)^2 = 1.256\,\text{Am}^2$$

$$A_{\text{Dipole}} = \frac{\mu_0}{4\pi}\frac{m\sin\theta}{r^2}$$

Substitute $\theta = 45^0, r = 1$ m

$$A_{\text{Dipole}} = \frac{4\pi \times 10^{-7} \times 1.256 \times \sin 45^0}{4\pi \times (1)^2} = 8.8 \times 10^{-8}\,\text{Tm}$$

Example 3.18 Consider a square current loop with side length $a = 0.1$ m and current $I = 5$ A, positioned such that its centre is at the origin. Calculate the vector potential $\vec{A}$ at a distance $r = 2$ m along the axis perpendicular to the plane of the loop and quadrupole terms.

Solution:

Since the loop is square with side a, area $A = a^2 = 0.01\,\text{m}^2$

$$m = IA = 0.05\,\text{Am}^2$$

$$A_{\text{Dipole}} = \frac{\mu_0}{4\pi}\frac{m}{r^2} = 1.25 \times 10^{-9}\,\text{Tm}$$

The quadrupole term for $\theta = 0^0$ is given by

$$A_{\text{quadrupole}} = \frac{\mu_0}{4\pi}\frac{Ia^2\cos 2\theta}{r^3} = 6.25 \times 10^{-10}\,\text{Tm}$$

Total vector potential

$$A = A_{\text{Dipole}} + A_{\text{quadrupole}} = 1.875 \times 10^{-9} \, \text{Tm}$$

Unsolved Problems:

Problem 3.1 A thunder cloud is stationary above level ground. Regarding the earth as a perfect conductor and the thunder cloud as an electric dipole with its axis vertical, show that the electric fieldat a point on the ground is proportional to $3\sin^5\alpha - \sin^3\alpha$, where α is the elevation of the cloud from the point.

Problem 3.2 A point charge is placed between two semi-infinite conducting plates which are inclined at angle of 30^0 with respect to each other. Calculate the number of image charges.

Ans. 11.

Problem 3.3 An infinitely long thin cylindrical shell has its axis coinciding with z-axis. It carries asurface charge density $\sigma_0\cos\phi$, where ϕ is the polar angle and σ_0 is constant. Calculate the magnitude of the electric field inside the cylinder.

Ans. $\frac{\sigma_0}{2\epsilon_0}$.

Problem 3.4 Calculate the electrostatic charge density $\rho(\vec{r})$ corresponding to the potential $V(r) = \frac{q}{4\pi\epsilon_0}\frac{1}{r}\left(1 + \frac{\alpha r}{2}\right)exp(-\alpha r)$

Ans. $-\frac{q}{4\pi}\alpha^3\frac{e^{-\alpha r}}{2}$.

Problem 3.5 A grounded conducting sphere of radius a is placed with its centre at the origin. A point dipole of dipole moment $\vec{p} = p\hat{k}$ is placed at a distance d along the x-axis, where $\hat{i}, \hat{j}$ are the unit vectors along the x and z-axes respectively. This leads to the formation of an image dipole of strength $\vec{p}$ at a distance d' from the centre along the x-axis. If $d' = \frac{a^2}{d}$, then calculate $\vec{p}'$.

Ans. $\vec{p}' = -\frac{a^3}{d^3}p\hat{k}$.

Problem 3.6 A grounded metal sheet is located in the $z = 0$ plane, while a point charge Q is located at $(0, 0, a)$. Find the force acting on a point charge -Q placed at $(a, 0, a)$.

Ans. $\frac{Q^2}{4\pi\epsilon_0 a^2}(-9.1\hat{a}_x - 0.071\hat{a}_y)N$.

Problem 3.7 A point charge $+q$ is placed at $(0, 0, d)$ above a grounded infinite conducting plane defined by $z = 0$. There are no charges present anywhere else. What is the magnitude of the electric field at $(0, 0, -d)$?

Ans. $\frac{q}{16\pi\epsilon_0 d^2}$.

Problem 3.8 Calculate the magnetic field corresponding to the vector potential $\vec{A} = \frac{1}{2}\vec{F} \times \vec{r} + \frac{10}{r^3}\vec{r}$, where $\vec{F}$ is a constant vector.

Ans. $\vec{F}$.

Problem 3.9 A charge distribution has a charge density given by $\rho = Q\{\delta(x - x_0) - \delta(x + x_0)\}$. For this charge distribution calculate the electric field at $(2x_0, 0, 0)$

  Ans. $\dfrac{Q}{8\pi\epsilon_0 x_0^2}\hat{x}$.

Problem 3.10 A charged particle is at a distance d from an infinite conducting plane maintained at zero potential. When released from rest, the particle reaches a speed u at a distance $\frac{d}{2}$ from the plane. At what distance from the plane will the particle reach the speed $2u$?

  Ans. $\frac{d}{5}$.

Problem 3.11 Consider an axially symmetric charge distribution of the form, $\rho = \rho_0\left(\frac{r_0}{r}\right)^2 e^{-\frac{r}{r_0}}\cos^2\phi$. Calculate the radial component of the dipole moment due to this charge distribution.

  Ans. $2\pi\rho_0 r_0^{\,3}$.

Problem 3.12 Two-point charges $+3Q$ and $-Q$ are placed at $(0, 0, d)$ and $(0, 0, 2d)$ respectively, above an infinite grounded conducting sheet kept in xy-plane. At a point $(0, 0, z)$, where $z \gg d$, calculate the approximate electrostatic potential of this charge configuration.

  Ans. $\dfrac{1}{4\pi\epsilon_0}\dfrac{2d}{z^2}Q$.

3.8 Summary

- **The Chapter Introduces Three Methods**: Separation of Variables, Method of Images, and Finite Element Analysis (for 2D cases).
- **Method of Images Concept**: Imaginary (image charges) are used to simplify boundary problems by meeting boundary conditions. These charges are positioned outside the region of interest to emulate the influence of actual charges near conductive surfaces.

  - We have derived the potential due to a real charge and its image charge.
  - We have derived the formulas for electric field, surface charge density, Coulomb force and work required to move charges.

- **Electrostatic Interaction with a Grounded Conducting Sphere**: The behaviour of a source charge is near a grounded sphere by introducing image charges. We have made calculations for potential and force due to interactions between the charge and the sphere.
- **Special Cases for Force Calculation**:

  - Short Distance: Force approximates Coulomb's law.
  - Long Distance: Force diminishes proportionally to the cube of separation, indicating a deviation from Coulomb's law at larger distances.

- **Surface Charge Density on Conducting Sphere**: Formula for induced surface charge density is derived and analyzed for different positions (angles) on the sphere's surface. It demonstrates the variation of charge density with angle θ.
- **Point Charge Near an Insulated Conducting Sphere**:

  - Analyzes the potential and surface charge distribution when a point charge interacts with an insulated sphere carrying a charge Q.
  - Uses the superposition principle to calculate the potential and force on the charge.

- **Conducting Sphere in a Uniform Electric Field**:

  - Examines the potential around a conducting sphere placed in a uniform electric field.
  - Introduces the image charges required to meet boundary conditions and calculates the induced surface charge density.

- **Multipole Expansion and Electric Dipole Potential**:

  - Explains the multipole expansion for the electric potential due to complex charge distributions.
  - Covers monopole, dipole and quadrupole terms, and the behaviour of electric potential at large distances.

- **Vector Potential Using Multipole Moments**:

  - Introduces the concept of vector potential, which involves expanding terms in powers of $1/r$.
  - Discusses the magnetic dipole moment and its independence from the origin, relating to Maxwell's equation.

- **Conducting Sphere in a Uniform Electric Field**:

  - Placing a sphere in a uniform electric field generates image charges that balance the field at the sphere's surface.
  - Potential and surface charge density for different regions on the sphere's surface are calculated.

- **Multipole Expansion and Dipole Moment**:

  - Multipole expansion is applied to complex charge distributions, breaking down potential into monopole, dipole and quadrupole components.
  - Provides a deeper understanding of electric potential at different distances, especially for systems with symmetrical charge distributions.

- **Vector Potential and Magnetic Dipole**:

  - Vector potential is expanded using multipole moments to represent magnetic fields for current-carrying loops.

– The concept of magnetic dipole moment is introduced, noting its consistency with Maxwell's equations, specifically $\vec{\nabla}.\vec{B} = 0$.

- **Applications of Dipole Potential and Magnetic Fields**:

  – Electric potential for dipoles and quadrupoles is calculated, showing behaviour at distances.
  – Magnetic dipole configurations, vector potentials and field calculations reinforce understanding of magnetic multipoles and their influence on nearby fields.

- **Examples and Unsolved Problems**:

  – Includes solved examples illustrating applications of the concepts, such as calculating force, potential and surface charge density for different configurations.
  – Provides unsolved problems to practice calculations based on learned principles.

Chapter 4
Dynamics of Electric and Magnetic Fields

Abstract This chapter explores the fundamentals and advanced formulations of magnetic phenomena, starting with the transition from electrostatic to magnetostatic fields. Magnetic fields, arising from currents, are distinguished from electric fields, which originate from charges. Utilizing the vector potential $\vec{A}$ via Biot-Savart law provides a practical and theoretical framework, analogous to the electrostatic potential V. Maxwell's equations integrate time-varying electric and magnetic fields, redefining them in terms of potentials. Gauge transformations reveal the flexibility in electrodynamic formulations, with the Coulomb gauge simplifying static problems and the Lorentz gauge providing symmetry for relativistic contexts. Continuous charge distributions and retarded potentials account for finite electromagnetic propagation speeds. Jefimenko's equations and Lineard–Wiechert potentials describe fields influenced by dynamic and moving sources, elucidating velocity- and acceleration-dependent effects. The fields of a moving point charge are analyzed, with components reflecting both Coulombic and radiative influences. This comprehensive approach bridges theory and application, enhancing understanding of magnetic fields, vector potentials and their electrodynamic implications.

Keywords Biot-Savart law · Gauge transformations · Jefimenko's equations · Lineard–Wiechert potentials

4.1 Introduction

So far we have discussed various electrostatic phenomena and now we turn to study steady-state magnetic phenomena. It is pertinent to mention here that the basic laws of magnetic fields did not follow directly from human being's earliest contact with magnetic materials. There may be several reasons but they mainly follow from the radical difference between magnetostatics and electrostatics. There are no free magnetic charges although the notation of a magnetic charge density may be a useful mathematical construct in certain circumstances. As discussed already, we are more comfortable if we compute V instead of computing $\vec{E}$. The advantage of working

R. Ahmad Parra et al., *A Systematic Approach to Electrodynamics*,
https://doi.org/10.1007/978-981-96-3408-8_4

with V is that it is scalar, nonetheless $\vec{E}$ is vector. On the other hand, in electro-dynamics $\vec{E} = -\vec{\nabla}V$ will not work. In the same way in magnetostatics, we can have a potential rather than field. Potential formulation is important for theoretical aspect/understanding. However, for practical purposes, it is better to deal with the magnetic field. In case of electrostatics, we measure electric potential, but in lab we can't measure vector potential. In electrostatic we measure potential difference not absolute potential. For static magnetic field, there exists a potential that we can under-stand mathematically. Experimentally it is confirmed that current-carrying conductor is having magnetic field around it. Current density is the moving charge density.

4.2 Vector Potential

The Coulomb's law and the Biot-Savart laws are, respectively, as:

$$\vec{E} = \frac{1}{4\pi\epsilon_0} \int \frac{\rho(\vec{r}')\hat{R}}{R^2}d\tau' \tag{4.1}$$

And

$$\vec{B} = \frac{\mu_0}{4\pi} \int \vec{J}(\vec{r}') \times \frac{\hat{R}}{R^2}d\tau' \tag{4.2}$$

From the above equations it is evident that, we need source $\rho(\vec{r}')$ to create $\vec{E}$ and source $\vec{J}(\vec{r}')$ to create $\vec{B}$. Only difference between $\vec{E}$ and $\vec{B}$ is the source.

Remember

$$\vec{\nabla}(1/R) = \frac{-\hat{R}}{R^2}; \quad \text{where } \vec{R} = \vec{r} - \vec{r}'$$

$$\vec{\nabla}'(1/R) = -\vec{\nabla}(1/R) \quad \text{and} \quad \nabla^2(1/R) = -4\pi\delta^3(\vec{r} - \vec{r}')$$

We employ following formulae in order to develop relations for magnetic fields

$$\vec{\nabla} \times \left(f\vec{A}\right) = f\left(\vec{\nabla} \times \vec{A}\right) - \vec{A} \times \left(\vec{\nabla}f\right)$$

and

$$\vec{\nabla} \cdot \left(f\vec{A}\right) = f\left(\vec{\nabla} \cdot \vec{A}\right) + \vec{A} \cdot \left(\vec{\nabla}f\right)$$

In our case $\vec{J} \to \vec{A}$ and $f \to \frac{1}{r}$

Equation (4.2) can, therefore, be written as

$$\vec{B}(\vec{r}) = \frac{-\mu_0}{4\pi} \int \vec{J}(\vec{r}') \times \vec{\nabla}(1/R)\mathrm{d}\tau'$$

(4.3)

which is simplified as follows:

$$\vec{B}(\vec{r}) = \int \frac{\mu_0}{4\pi}\left[\vec{\nabla} \times \left(\frac{\vec{J}(\vec{r}')}{R}\right) - \frac{1}{R}\left(\vec{\nabla} \times \vec{J}(\vec{r}')\right)\right]\mathrm{d}\tau'$$

$$\vec{B}(\vec{r}) = \int \frac{\mu_0}{4\pi}\left[\vec{\nabla} \times \left(\frac{\vec{J}(\vec{r}')}{R}\right)\right]\mathrm{d}\tau'$$

$$\vec{B}(\vec{r}) = \frac{\mu_0}{4\pi}\left[\vec{\nabla} \times \int \frac{\vec{J}(\vec{r}')}{|\vec{r}-\vec{r}'|}\mathrm{d}\tau'\right]$$

(4.4)

Here, we have changed the order of integral and $\vec{\nabla}$ operates on x, y, z and integral operates on x', y', z'.

$$\vec{B}(\vec{r}) = \vec{\nabla} \times \left[\frac{\mu_0}{4\pi}\int \frac{\vec{J}(\vec{r}')}{|\vec{r}-\vec{r}'|}\mathrm{d}\tau'\right]$$

(4.5)

The quantity within the brackets is a vector quantity, symbolized as $\vec{A}$. Hence, we can write the magnetic field as

$$\vec{B}(\vec{r}) = \left(\vec{\nabla} \times \vec{A}\right)$$

(4.6)

where the quantity $\vec{A}$ is defined as

$$\vec{A} = \frac{\mu_0}{4\pi}\int \frac{\vec{J}(\vec{r}')}{|\vec{r}-\vec{r}'|}\mathrm{d}\tau'$$

(4.7)

Conclusion: It is evident from the above illustration that we can deduce magnetic field, provided we know another vector quantity $\vec{A}$, reckoned as the magnetic vector potential. Further, this concept can be generalized to solve Gauss law, $\vec{\nabla} \cdot \vec{B}.\vec{\nabla} \cdot \vec{B}$.

$$\vec{\nabla} \cdot \vec{B} = \vec{\nabla} \cdot \left(\vec{\nabla} \times \vec{A}\right)$$

We know that the divergence of curl vanishes always. Hence, we can write

$$\vec{\nabla} \cdot \vec{B} = 0$$

(4.8)

which is dubbed as Gauss law and it signifies that the magnetic monopoles do not exist at all. This equation also permits us to express $\vec{B}$ as the curl of some vector field $\vec{A}$, called the vector potential.

$$\vec{\nabla} \times \vec{B} = \vec{\nabla} \times \left(\vec{\nabla} \times \vec{A} \right)$$

$$\vec{\nabla} \times \vec{B} = \vec{\nabla} \left(\vec{\nabla} \cdot \vec{A} \right) - \nabla^2 \vec{A}$$

The above equation can be simplified if we insert the notation of magnetic vector potential

$$\vec{\nabla} \times \vec{B} = \frac{\mu_0}{4\pi} \left[\vec{\nabla} \left(\vec{\nabla} \cdot \int \frac{\vec{J}\left(\vec{r}'\right)}{|\vec{r} - \vec{r}'|} d\tau' \right) \right] - \left[\frac{\mu_0}{4\pi} \nabla^2 \int \frac{\vec{J}\left(\vec{r}'\right)}{|\vec{r} - \vec{r}'|} d\tau' \right] \qquad (4.9)$$

Again, the same argument $\vec{\nabla}$ operates on x, y, z and integral operates on x', y', z'. Let

$$I_1 = \frac{\mu_0}{4\pi} \left[\vec{\nabla} \left(\vec{\nabla} \cdot \int \frac{\vec{J}\left(\vec{r}'\right)}{|\vec{r} - \vec{r}'|} d\tau' \right) \right] \qquad (4.10)$$

which can be further modified as

$$I_1 = \frac{\mu_0}{4\pi} \left[\vec{\nabla} \left(\vec{\nabla} \cdot \int \frac{\vec{J}\left(\vec{r}'\right)}{R} d\tau' \right) \right] \qquad (4.11)$$

which resembles with the following integral

$$\int \vec{\nabla} \cdot \left(K\vec{F} \right) d\tau', \quad \text{where } \vec{J}\left(\vec{r}'\right) = \vec{F} \text{ and } K = \frac{1}{R}$$

From the Gauss Divergence theorem, we know that

$$\int \left(\vec{\nabla} \cdot \vec{V} \right) d\tau' = \oint \vec{V} \cdot d\vec{s}'$$

Therefore, we can write

$$\int \left(\vec{\nabla} \cdot \left(K\vec{F} \right) \right) d\tau' = \oint \left(K\vec{F} \right) \cdot d\vec{s}'$$

Using the same concept in our problem, we can write as

$$\vec{\nabla} \cdot \int \frac{\vec{J}(\vec{r}\prime)}{R} \mathrm{d}\tau' = \oint \frac{\vec{J}(\vec{r}') \cdot \mathrm{d}\vec{s}'}{R} = 0 \tag{4.12}$$

Current flowing normal to the surface, i.e., $\int \vec{J}(\vec{r}') \cdot \mathrm{d}\vec{s}'$, is the current flowing normal to surface. R.H.S is the algebraic sum of currents in a closed loop, that should be (zero). Hence, we can write $I_1 = 0$.

Next, we consider I_2 as follows:

$$I_2 = \frac{\mu_0}{4\pi} \nabla^2 \int \frac{\vec{J}(\vec{r}')}{|\vec{r} - \vec{r}'|} \mathrm{d}\tau' \tag{4.13}$$

$$I_2 = \frac{\mu_0}{4\pi} \int \vec{J}(\vec{r}') \nabla^2 \left(\frac{1}{R}\right) \mathrm{d}\tau'$$

$$I_2 = \frac{\mu_0}{4\pi} \int \vec{J}(\vec{r}') \cdot \left(-4\pi \delta^3 (\vec{r} - \vec{r}')\right) \mathrm{d}\tau'$$

$$I_2 = -\mu_0 \int \vec{J}(\vec{r}') \cdot \left(\delta^3 (\vec{r} - \vec{r}')\right) \mathrm{d}\tau'$$

$$I_2 = -\mu_0 \vec{J}(\vec{r}') \tag{4.14}$$

Hence Eq. (4.9) becomes

$$\vec{\nabla} \times \vec{B} = I_1 - I_2 = 0 - \left(-\mu_0 \vec{J}(\vec{r}')\right)$$

$$\vec{\nabla} \times \vec{B} = \mu_0 \vec{J}(\vec{r}') \tag{4.15}$$

which is only valid for magnetostatics as we have assumed that uniform current density is present in the circuit. In case of non-steady current $I_1 \neq 0$. We see that as we started from the Biot-Savart law, we have established the following.

$$\vec{\nabla} \cdot \vec{B} = 0$$
$$\vec{B} = \vec{\nabla} \times \vec{A}$$

and

$$\vec{\nabla} \times \vec{B} = \mu_0 \vec{J}(\vec{r}')$$

4.3 Potentials and Fields Formulations

Maxwell's relations discuss the behaviour of electromagnetic fields. These are a set of coupled first-order partial differential equations which relate electric and magnetic fields. These relations can be enumerated as follows:

$$\vec{\nabla} \cdot \vec{E} = \frac{\rho}{\epsilon_0}$$

$$\vec{\nabla} \times \vec{E} = \frac{-\partial \vec{B}}{\partial t}$$

$$\vec{\nabla} \cdot \vec{B} = 0$$

$$\vec{\nabla} \times \vec{B} = \mu_0 \vec{J} + \mu_0 \epsilon_0 \frac{\partial \vec{E}}{\partial t} \tag{4.16}$$

Here $\rho(\vec{r}, t)$ and $\vec{J}(\vec{r}, t)$ are the given sources. We can deduce the fields $\vec{E}(\vec{r}, t)$ and $\vec{B}(\vec{r}, t)$ corresponding to $\rho(\vec{r}, t)$ and $\vec{J}(\vec{r}, t)$, provided we utilized the concept of Coulomb's law and the Biot-Savart law in static case. We aim to generalize these laws to the time-dependent charge configuration. In electrostatics $\vec{\nabla} \times \vec{E} = 0$, which can be extended to write electric field, $\vec{E}$ as the negative gradient of scalar potential. However, this concept cannot be extended to electrodynamics because the curl of $\vec{E}$ does not vanish there. But $\vec{B}$ remains divergenceless. Therefore, we can write

$$\vec{\nabla} \times \vec{E} = \frac{-\partial \vec{B}}{\partial t} \quad \text{and} \quad \vec{B} = \vec{\nabla} \times \vec{A}$$

Thus, the Eq. (4.16) attains the following form

$$\vec{\nabla} \times \vec{E} = \frac{-\partial \left(\vec{\nabla} \times \vec{A} \right)}{\partial t}$$

$$\vec{\nabla} \times \left(\vec{E} + \frac{\partial \vec{A}}{\partial t} \right) = 0 \tag{4.17}$$

It is noteworthy that if the curl of some physical quantity vanishes, it can, therefore, be explicitly exhibited as the gradient of some scalar potential function, i.e.,

$$\vec{E} + \frac{\partial \vec{A}}{\partial t} = -\vec{\nabla} V \tag{4.18}$$

From this expression it follows that

$$\vec{E} = -\vec{\nabla} V - \frac{\partial \vec{A}}{\partial t} \tag{4.19}$$

Substitute Eq. (4.19) in Maxwell's first equation, we get

$$\nabla^2 V + \frac{\partial\left(\vec{\nabla}\cdot\vec{A}\right)}{\partial t} = -\frac{\rho}{\epsilon_0} \tag{4.20}$$

which reduces to Poisson's equation in the steady-state case. Putting the above equations in the Maxwell's fourth equation, we get

$$\vec{\nabla}\times\left(\vec{\nabla}\times\vec{A}\right) = \mu_0\vec{J} - \mu_0\epsilon_0\vec{\nabla}\frac{\partial V}{\partial t} - \mu_0\epsilon_0\frac{\partial^2\vec{A}}{\partial t^2}$$

Using the identity, $\vec{\nabla}\times\left(\vec{\nabla}\times\vec{A}\right) = \vec{\nabla}\left(\vec{\nabla}.\vec{A}\right) - \nabla^2\vec{A}$ and rearranging the terms, we get

$$\left(\nabla^2\vec{A} - \mu_0\epsilon_0\frac{\partial^2\vec{A}}{\partial t^2}\right) - \vec{\nabla}\left(\vec{\nabla}\cdot\vec{A} + \mu_0\epsilon_0\frac{\partial V}{\partial t}\right) = -\mu_0\vec{J}$$

$$\vec{\nabla}\left(\vec{\nabla}\cdot\vec{A}\right) - \nabla^2\vec{A} = \mu_0\vec{J} - \mu_0\epsilon_0\vec{\nabla}\frac{\partial V}{\partial t} - \mu_0\epsilon_0\frac{\partial^2\vec{A}}{\partial t^2}$$

$$\nabla^2\vec{A} - \mu_0\epsilon_0\frac{\partial^2\vec{A}}{\partial t^2} - \vec{\nabla}\left(\vec{\nabla}\cdot\vec{A} + \mu_0\epsilon_0\frac{\partial V}{\partial t}\right) = -\mu_0\vec{J} \tag{4.21}$$

Thus, the set of four-Maxwell's relations have been reduced to two coupled differential equations.

Example 4.1 A finite wire segment of length $L = 1.0$ m carrying current $I = 8$ A lies along the z-axis from $z = -0.5$ m to $z = +0.5$ m. Calculate the vector potential A at a point 0.6 m from the wire along the x-axis.

Solution:

The vector potential at a perpendicular distance r from the wire segment is given by

$$A_\phi = \frac{\mu_0 I}{4\pi}\int_{L/2}^{L/2}\frac{1}{\sqrt{r^2+z^2}}dz$$

$$A_\phi = \frac{\mu_0 I}{4\pi}\left[\ln\left(z+\sqrt{r^2+z^2}\right)\right]\Big|_{-0.5}^{+0.5}$$

$$A_\phi = 1.21\times10^{-6}\ \text{Tm}$$

Example 4.2 Consider a current density $\vec{J}(x,y,z,t) = J_0\sin(\omega t)\delta(x)\delta(y)\delta(z)\hat{z}$, where J_0 is a constant and ω is the angular frequency. Determine the electric field $\vec{E}$ generated by this time-dependent current source.

Solution:

$\vec{A} = \frac{\mu_0}{4\pi} \int \frac{J(x',y',z',t-R/c)}{R} d^3x'$, we can calculate $\vec{A}$ directly as

$$\vec{A}(r, t) = \frac{\mu_0 J_0 \sin(\omega(t - R/c))}{4\pi R} \hat{z}$$

The electric field $\vec{E}$ is $\vec{E} = -\vec{\nabla}V - \frac{\partial \vec{A}}{\partial t}$
Assuming $V = 0$

$$\vec{E} = -\frac{\mu_0 J_0 \omega \cos(\omega(t - R/c))}{4\pi R} \hat{z}$$

Example 4.3 An oscillating charge distribution is given by $\rho(x, y, z, t) = \rho_0 e^{-r^2/a^2}$, where ρ_0 and a are constants, and ω is the angular frequency. Find the vector potential $\vec{A}$ generated by this charge distribution.

Solution:

The current density $\vec{J} = \rho\vec{v}$, assuming a radial oscillation $\vec{v} = v_0 \sin(\omega t)\hat{r}$

$$\vec{J}(r, t) = \rho_0 e^{-r^2/a^2} v_0 \sin(\omega t)\hat{r}$$

The vector potential is given by

$$\vec{A}(r, t) = \frac{\mu_0}{4\pi} \int \frac{\vec{J}\left(\vec{r}', t - |r - \vec{r}'|\right) d^3\vec{r}'}{|r - \vec{r}'|}$$

Substitute $\vec{J}\left(\vec{r}', t\right)$ and assume $\vec{r} \gg \vec{r}' \vec{r} \gg \vec{r}'$

$$\vec{A}(r, t) = \frac{\mu_0 \rho_0 v_0 \pi^{3/2} a^3}{4\pi r} \sin(\omega(t - r/c))\hat{r}$$
$$\vec{A}(r, t) = \frac{\mu_0 \rho_0 v_0 \pi^{3/2} a^3}{4r} \sin(\omega(t - r/c))\hat{r}$$

Example 4.4 Given $V(\vec{r}, t) = V_0 \frac{\cos(\omega t)}{r}$ and $\vec{A}(r, t) = A_0 \frac{\sin(\omega t)}{r} \hat{r}$. Determine the electric field $\vec{E}$.

Solution:

Using $\vec{E} = -\vec{\nabla}V - \frac{\partial \vec{A}}{\partial t}$

$$\vec{\nabla}V = -V_0 \frac{\cos(\omega t)}{r^2} \hat{r}$$

$$\frac{\partial \vec{A}}{\partial t} = A_0 \frac{\omega \cos(\omega t)}{r} \hat{r}$$

$$\vec{E} = -\left(-V_0 \frac{\cos(\omega t)}{r^2} \hat{r}\right) - A_0 \frac{\omega \cos(\omega t)}{r} \hat{r}$$

$$\vec{E} = \left(\frac{V_0}{r^2} + \frac{\omega A_0}{r}\right) \cos(\omega t) \hat{r}$$

4.4 Gauge Transformations

We can deduce ρ and $\vec{J}$ provided we are familiar with fields $\vec{E}$ and $\vec{B}$. Equations (4.20) and (4.21) are coupled differential equations; however, we have been successfully able to reduce a six-body problem $\left(\vec{E}, \vec{B}\right)$ to four body problem $\left(V, \vec{A}\right)$. The expressions $\nabla^2 V + \frac{\partial\left(\vec{\nabla}.\vec{A}\right)}{\partial t} = -\frac{\rho}{\epsilon_0}$ and $\vec{B} = \vec{\nabla} \times \vec{A}$ do not uniquely determine the potentials. We are free to impose extra conditions on V and $\vec{A}$, as long as nothing happens to $\vec{E}$ and $\vec{B}$. This can be done by using gauge transformations.

Let $\left(V, \vec{A}\right)$ and $\left(V', \vec{A}'\right)$, be the two sets of potentials, corresponding to the same electric and magnetic fields. Using the gauge transformation.

$$\vec{A}\prime = \vec{A} + \vec{\alpha} \text{ and } V' = V + \beta \tag{4.22}$$

Since, the two $\vec{A}'s$ results in same $\vec{B}$, their curls must be equal, hence $\vec{\nabla} \times \vec{\alpha} = 0$. Therefore, we can illustrate $\vec{\alpha}$ as the gradient of some scalar potential as under.

$$\vec{\alpha} = \vec{\nabla}\lambda \tag{4.23}$$

Since, these two potentials result in the same electric field $\vec{E}$, therefore, we can write

$$\vec{\nabla}\beta + \frac{\partial \vec{\alpha}}{\partial t} = 0$$

$$\vec{\nabla}\left(\beta + \frac{\partial \lambda}{\partial t}\right) = 0 \tag{4.24}$$

The term within the brackets is, therefore, independent of position, however it could depend on time, we can call it as $K(t)$. Taking gradient of V' and $\vec{A}\prime\vec{A}\prime$

$$\vec{\nabla}V' = \vec{\nabla}V + \vec{\nabla}\beta \tag{4.25}$$

$$\frac{\partial \vec{A}'}{\partial t} = \frac{\partial \vec{A}}{\partial t} + \frac{\partial \vec{\alpha}}{\partial t} \tag{4.26}$$

Adding Eqs. (4.25) and (4.26) and multiplying the result by minus sign, we get

$$-\vec{\nabla} V' - \frac{\partial \vec{A}'}{\partial t} = -\vec{\nabla} V - \frac{\partial \vec{A}}{\partial t} - \vec{\nabla}\beta - \frac{\partial \vec{\alpha}}{\partial t}$$

$$\vec{E} = \vec{E} - \left(\vec{\nabla}\beta + \frac{\partial \vec{\alpha}}{\partial t} \right)$$

$$\vec{\nabla}\beta + \frac{\partial \vec{\alpha}}{\partial t} = 0$$

$$\beta = \frac{-\partial \lambda}{\partial t} + K(t)$$

$$K(t) = \beta + \frac{\partial \lambda}{\partial t} \tag{4.27}$$

We could, as well absorb $K(t)$ in to λ, if we redefine λ' as follows:

$$\lambda' \rightarrow \lambda + \int_0^t K(t)\mathrm{d}t$$

On adding $\int_0^t K(t)\mathrm{d}t$, to the previous expression, does not alter the gradient of λ. It just adds $K(t)$ to $\frac{\partial \lambda}{\partial t}$. Therefore, it follows from the elucidation that

$$\vec{A}\prime = \vec{A} + \vec{\nabla}\lambda \quad \text{and} \quad V' = V - \frac{\partial \lambda}{\partial t} \tag{4.28}$$

Conclusion:

For any old scalar function $\lambda(\vec{r}, t)$, we can add $\vec{\nabla}\lambda$ to $\vec{A}$ provided we at the same time subtract $\frac{\partial \lambda}{\partial t}$ from V. This will not change the basic structure of the physical quantities $\vec{E}$ and $\vec{B}$. These mathematical changes in V and $\vec{A}$ are referred as gauge transformations and the invariance of fields under such transformations is called gauge invariance. However, we choose $\vec{\nabla} \cdot \vec{A} = 0$, in magnetostatics. But, in case of electrodynamics the picture is not so simple. The most convenient gauge depends to some extent on the problem at hand. There are some famous gauge transformations, which we will discuss in the succeeding sections.

4.5 Coulomb's Gauge

So far, we see that for many potentials, we have same $\vec{E}$ and $\vec{B}$, therefore, we are at ease to add anything to potential. As in magnetostatics, we take $\vec{\nabla} \cdot \vec{A} = 0$, Eq. (4.20) can be written as:

$$\nabla^2 V = \frac{-\rho}{\epsilon_0} \tag{4.29}$$

which is the famous equation called as Poisson's equation and its solution is given by as under

$$V(\vec{r}, t) = \frac{1}{4\pi\epsilon_0} \int \frac{\rho(\vec{r}', t)\mathrm{d}\tau'}{R} \tag{4.30}$$

Here, the potential can be determined by the distribution of charges. The essence of the Coulomb's gauge lies in the fact that the scalar potential can be evaluated easily. The disadvantage is that the vector potential, $\vec{A}$ is extremely difficult to be calculated from the Coulomb's gauge. The differential equations for $\vec{A}$ in the Coulomb's gauge is written as under

$$\nabla^2 \vec{A} - \mu_0\epsilon_0 \frac{\partial^2 \vec{A}}{\partial t^2} = -\mu_0 \vec{J} + \mu_0\epsilon_0 \vec{\nabla}\left(\frac{\partial V}{\partial t}\right) \tag{4.31}$$

4.5.1 The Lorentz Gauge

Lorentz gauge condition is written as follows:

$$\vec{\nabla} \cdot \vec{A} = -\mu_0\epsilon_0 \frac{\partial V}{\partial t} \tag{4.32}$$

This gauge condition will help to uncouple the pair of Eqs. (4.20) and (4.21) and, thereby produces two inhomogeneous wave equations simultaneously for V and $\vec{A}$ as follows:

$$\nabla^2 V - \mu_0\epsilon_0 \frac{\partial^2 V}{\partial t^2} = \frac{-\rho}{\epsilon_0} \tag{4.33}$$

$$\nabla^2 \vec{A} - \mu_0\epsilon_0 \frac{\partial^2 \vec{A}}{\partial t^2} = -\mu_0 \vec{J} \tag{4.34}$$

The importance of the Lorentz gauge lies in the fact that it treats V and $\vec{A}$ on equivalent footings. Further, this concept is independent of the coordinate system chosen and therefore, fits easily into the considerations of special relativity. Further, we define d'Alembertian operator as follows:

$$\nabla^2 - \mu_0 \epsilon_0 \frac{\partial^2}{\partial t^2} = \Box^2 \tag{4.35}$$

Utilizing the concept of d'Alembertian operator, Eqs. (4.33) and (4.34) will respectively assume the following form

$$\Box^2 V = -\frac{\rho}{\epsilon_0} \tag{4.36}$$

$$\Box^2 \vec{A} = -\mu_0 \vec{J} \tag{4.37}$$

The delicate procedure of V and $\vec{A}$ is very awesome in the context of theory of special relativity, where, the d'Alembertian is the natural extension of Laplacian and Eqs. (4.36) and (4.37) can be referred as four-dimensional versions of Poisson's equation. However, in the Lorentz gauge V and $\vec{A}$ satisfy inhomogeneous wave equations where the source term is placed on the right side of the equation. Therefore, the entire theory of electrodynamics revolves around the problem of solving inhomogeneous wave equations for specified sources.

Example 4.5 Given an initial vector potential $\vec{A}(r, t) = A_0 e^{-(x^2+y^2)} \hat{z}$ and the scalar potential $V = V_0 x e^{-t}$, find the gauge transformation needed to make the scalar potential V' time independent, if possible.

Solution:

The gauge transformations are defined as

$$\vec{A}' = \vec{A} + \vec{\nabla}\lambda, \quad V' = V - \frac{\partial\lambda}{\partial t}$$

We aim V' to be time independent $\frac{\partial V'}{\partial t} = 0 \Rightarrow \frac{\partial^2\lambda}{\partial t^2} = \frac{\partial V}{\partial t} = -V_0 x e^{-t}$
With $V = V_0 x e^{-t}$

$$\frac{\partial\lambda}{\partial t} = V_0 x e^{-t}$$

On integrating, we find $\frac{\partial\lambda}{\partial t} = -V_0 x e^{-t} + g(x, y, z)$

Using this in the above equation gauge transformation results in V' being independent of time as desired.

Example 4.6 Given potentials $\vec{A}(\vec{x}, t) = A_0 cos(kx - \omega t)\hat{x}$ and $V(\vec{x}, t) = V_0 \sin(kx - \omega t)$. Find the gauge function λ to transform $\vec{A}$ and V into the Coulomb gauge, where $\vec{\nabla} \cdot \vec{A}' = 0$.

Solution:

The coulomb gauge condition requires $\vec{\nabla}.\vec{A}' = 0$ with $\vec{A}' = \vec{A} + \vec{\nabla}\lambda$.

We have

$$\vec{\nabla} \cdot \vec{A}' = \vec{\nabla}.\vec{A} + \nabla^2\lambda = 0$$
$$\vec{\nabla} \cdot \vec{A} = -kA_0 \sin(kx - \omega t)$$

Solving for $\lambda(\vec{x}, t)$, we get $\lambda(\vec{x}, t) = -\frac{A_0}{k} \sin(kx - \omega t)$

Hence $\vec{\nabla} \cdot \vec{A}' = 0$.

Example 4.7 For potentials $\vec{A}(\vec{x}, t) = A_0 \sin(kx)e^{-\alpha t}\hat{x}$ and $V(\vec{x}, t) = V_0 \cos(kx)e^{-\alpha t}$. Find a gauge transformation to ensure that the gauge transformation condition is $\vec{\nabla} \cdot \vec{A} + \frac{1}{c^2}\frac{\partial V}{\partial t} = 0$. Given that $k^2 = \frac{\alpha^2}{c^2}$

Solution:

The Lorentz gauge requires

$$\vec{\nabla} \cdot \vec{A} + \frac{1}{c^2}\frac{\partial V}{\partial t} = 0$$

$$\vec{\nabla} \cdot \vec{A} = kA_0 \cos(kx)e^{-\alpha t} \quad \text{and} \quad \frac{\partial V}{\partial t} = -\alpha V_0 \cos(kx)e^{-\alpha t}$$

Substituting into the gauge condition

$$kA_0 \cos(kx)e^{-\alpha t} - \frac{1}{c^2}\alpha V_0 \cos(kx)e^{-\alpha t} = 0$$

$$kA_0 = \frac{\alpha^2}{c^2}V_0$$

Assume a gauge function $\lambda(\vec{x}, t)$ such that

$$\vec{\nabla} \cdot \left(\vec{A} + \vec{\nabla}\lambda\right) + \frac{1}{c^2}\frac{\partial}{\partial t}\left(V - \frac{\partial \lambda}{\partial t}\right) = 0$$

For Lorentz condition to be met $\lambda(\vec{x}, t) = \frac{A_0}{k} \sin(kx - \omega t)e^{-\alpha t}$

$$\vec{\nabla} \cdot \vec{A}' + \frac{1}{c^2} \frac{\partial V'}{\partial t} = 0.$$

Example 4.8 Suppose $V = 0$ and $\vec{A} = A_0 \sin(kx - \omega t)\hat{y}$, where A_0, ω and k are constants. Find $\vec{E}$ and $\vec{B}$.

Solution:

The electric field is given by:

$$\vec{E} = -\vec{\nabla}\vec{V} - \frac{\partial \vec{A}}{\partial t}$$

$$\vec{E} = A_0 \, \omega \cos(kx - \omega t)\hat{y}$$

This shows that the electric field oscillates in the y-direction with amplitude $A_0\omega$ and the same spatial and temporal dependence as the vector potential.

The magnetic field is given by the curl of the vector potential: $\vec{B} = \vec{\nabla} \times \vec{A}$

$$\vec{B} = A_0 k \cos(kx - \omega t)\hat{z}$$

This shows that the magnetic field oscillates in the z-direction with amplitude $A_0 k$ and the same spatial and temporal dependence as the vector potential.

4.6   Continuous Charge Distributions

In the continuous charge distributions, we discuss the concept of retarded potentials.

4.6.1   Retarded Potentials

For the static case, we know that the equations for the scalar and vector potential are $\nabla^2 V = \frac{-\rho}{\epsilon_0}$ and $\nabla^2 \vec{A} = -\mu_0 \vec{J}$ respectively. We already know the solutions of these equations as:

$$V(\vec{r}) = \frac{1}{4\pi\epsilon_0} \int \frac{\rho(\vec{r}')\mathrm{d}\tau'}{R} \tag{4.38}$$

and

$$\vec{A}(\vec{r}) = \frac{\mu_0}{4\pi} \int \frac{\vec{J}(\vec{r}', t)\mathrm{d}\tau'}{R} \tag{4.39}$$

As, we know that the electromagnetic waves travel with the speed of light. Therefore, in non-static case it is not the state of the source that is of prime importance at some particular instant, but rather its condition at some earlier time t_r (retarded time) when the message has left.

$$t_r = t - \frac{R}{c} \tag{4.40}$$

where c is the speed of light in vacuum.

For the generalization of the non-static case, we can write Eqs. (4.38) and (4.39) as follows:

$$V(\vec{r}, t) = \frac{1}{4\pi \epsilon_0} \int \frac{\rho(\vec{r}', t_r)d\tau'}{R} \tag{4.41}$$

$$\vec{A}(\vec{r}, t) = \frac{\mu_0}{4\pi} \int \frac{\vec{J}(\vec{r}', t_r)d\tau'}{R} \tag{4.42}$$

These are the potentials that have to be calculated at a retarded time and are therefore, called as retarded potentials. However, it is logically correct that we did not solve Eqs. (4.38) and (4.39) but introduced an interesting argument that the electromagnetic waves traverse through vacuum with the speed of light. Thus, to prove Eqs. (4.41) and (4.42) are solutions, it is incumbent that these equations must satisfy inhomogeneous Eqs. (4.38) and (4.39) while conforming Lorentz conditions. Further, it may sound bizarre but it is logically sound argument that the same procedure cannot be applied to fields, i.e.,

$$\vec{E}(\vec{r}, t) \neq \frac{1}{4\pi \epsilon_0} \int \frac{\rho(\vec{r}', t_r)\hat{R}d\tau'}{R^2}$$

$$\vec{B}(\vec{r}, t) \neq \frac{\mu_0}{4\pi} \int \frac{\vec{J}(\vec{r}', t_r)\hat{R}d\tau'}{R^2}$$

$$R = |\vec{r} - \vec{r}'|$$

and

$$t_r = t - \frac{R}{c}$$

From Eq. (4.41), we can write:

$$\vec{\nabla}V = \frac{1}{4\pi \epsilon_0} \int \vec{\nabla}\left(\frac{\rho(\vec{r}', t_r)d\tau'}{R}\right)$$

$$\vec{\nabla} V = \frac{1}{4\pi \epsilon_0} \int \left[(\vec{\nabla}\rho) \frac{1}{R} + \rho \vec{\nabla}\left(\frac{1}{R}\right) \right] d\tau'$$

since

$$\vec{\nabla}\rho = \frac{\partial \rho}{\partial t_r} \frac{\partial t_r}{\partial r} = \dot{\rho} \vec{\nabla} t_r = \frac{-1}{c} \dot{\rho} \vec{\nabla} R$$

$$\vec{\nabla} R = \hat{R} \quad \text{and} \quad \vec{\nabla}\left(\frac{1}{R}\right) = \frac{-\hat{R}}{R^2}$$

$$\vec{\nabla} V = \frac{1}{4\pi \epsilon_0} \int \left[\frac{-\dot{\rho}}{c} \frac{\hat{R}}{R} - \rho \frac{\hat{R}}{R^2} \right] d\tau'$$

Taking divergence of above equation and noting the following fact, we get:

$$\vec{\nabla} \cdot \left(f\vec{A} \right) = f \left(\vec{\nabla} \cdot \vec{A} \right) + \vec{A} \cdot \left(\vec{\nabla} f \right)$$

$$\nabla^2 V = \frac{1}{4\pi \epsilon_0} \int \left[\frac{-1}{c} \vec{\nabla} \cdot \left(\dot{\rho} \frac{\hat{R}}{R} \right) - \vec{\nabla}\left(\rho \frac{\hat{R}}{R^2} \right) \right] d\tau'$$

$$\nabla^2 V = \frac{1}{4\pi \epsilon_0} \int \left[\frac{-1}{c} \left(\frac{\hat{R}}{R} \cdot \vec{\nabla}\dot{\rho} \right) + \dot{\rho} \vec{\nabla} \cdot \left(\frac{\hat{R}}{R} \right) - \left\{ \frac{\hat{R}}{R^2} \cdot \left(\vec{\nabla}\rho \right) + \rho \vec{\nabla}\left(\frac{\hat{R}}{R^2} \right) \right\} \right] d\tau'$$

$$(4.43)$$

But we know that:

$$\vec{\nabla}\dot{\rho} = \frac{-1}{c} \ddot{\rho} \vec{\nabla} R = \frac{-1}{c} \ddot{\rho} \hat{R} \quad \text{and} \quad \vec{\nabla} \cdot \left(\frac{\hat{R}}{R} \right) = \frac{1}{R^2}; \ \vec{\nabla} \cdot \left(\frac{\hat{R}}{R^2} \right) = 4\pi \delta^3 (\vec{r})$$

We incorporate above equations in Eq. (4.43) and the equation for $\nabla^2 V$ becomes

$$\nabla^2 V = \frac{1}{4\pi \varepsilon_0} \int \left[\frac{-1}{c} \left(\frac{\hat{R}}{R} \cdot \left(\frac{-1}{c} \ddot{\rho} \hat{R} \right) \right) + \dot{\rho} \frac{1}{R^2} \right.$$
$$\left. - \left\{ \frac{\hat{R}}{R^2} \cdot \left(\frac{-1}{c} \dot{\rho} \vec{\nabla} R \right) + 4\pi \rho \delta^3 (\vec{r}) \right\} \right] d\tau'$$

$$\nabla^2 V = \frac{1}{4\pi \varepsilon_0} \int \left[\frac{1}{c^2} \left(\frac{\ddot{\rho}}{R} \right) - 4\pi \rho \delta^3 (\vec{r}) \right] d\tau'$$

$$\nabla^2 V = \frac{1}{c^2} \frac{\partial^2 V}{\partial t^2} - \frac{1}{\varepsilon_0} \rho(\vec{r}, t)$$

$$\Box^2 V = \frac{-\rho}{\varepsilon_0}$$

Thus Eq. (4.41) satisfies inhomogeneous Eq. (4.36) and proceeding in the same manner, we can prove that Eq. (4.42) satisfies inhomogeneous Eq. (4.37). The same is true for the advanced potentials

$$V_{\mathrm{a}}(\vec{r}, t) = \frac{1}{4\pi\epsilon_0} \int \left[\left(\frac{\rho(r, t_{\mathrm{a}})}{R} \right) \right] \mathrm{d}\tau' \tag{4.44}$$

and

$$\vec{A}_{\mathrm{a}}(\vec{r}, t) = \frac{\mu_0}{4\pi} \int \left[\left(\frac{\vec{J}(r, t_{\mathrm{a}})}{R} \right) \right] \mathrm{d}\tau' \tag{4.45}$$

Here

$$t_{\mathrm{a}} = t + \frac{R}{c}$$

It is pertinent to mention here that the advanced potentials are very consistent with the Maxwell's equations, but they violate the principle of causality.

Example 4.9 A spherical volume of radius R centred at the origin has a uniform charge density $\rho(r', t) = \rho_0 \sin(\omega t)$, where ρ_0 is a constant. Find the potential due to a uniform distribution.

Solution:

Retarded time $t_{\mathrm{r}} = t - R/c$.

For continuous charge distribution scalar potential is

$$V(\vec{r}, t) = \frac{1}{4\pi\epsilon_0} \int \frac{\rho(\vec{r}, t_{\mathrm{r}})}{|\vec{r} - \vec{r}'|} \mathrm{d}\tau'$$

The potential due to a uniform distribution within a sphere is equivalent to a point charge at the origin. Hence

$$V(\vec{r}, t) = \frac{1}{4\pi\epsilon_0} \frac{Q(t_{\mathrm{r}})}{r}$$

where $Q(t_{\mathrm{r}}) = \rho_0 \sin(\omega t_{\mathrm{r}}) \frac{4}{3}\pi R^3$

$$V(\vec{r}, t) = \frac{1}{4\pi\epsilon_0} \frac{\rho_0 \sin(\omega(t - R/c)) \frac{4}{3}\pi R^3}{r}$$

$$V(\vec{r}, t) = \frac{\rho_0 R^3 \sin(\omega(t - R/c))}{3\epsilon_0 r}$$

4.6.2　Jefimenko's Equations

In case of static charge and current distributions, the solutions of electric and magnetic fields are expressed in guise of Coulomb's law and Biot-Savart law. However, for time-dependent charge and current distributions, the solutions for the fields are known as Jefimenko's equations. We have the retarded potentials as

$$V(\vec{r}, t) = \frac{1}{4\pi\epsilon_0} \int \left[\left(\frac{\rho(r, t_r)}{R} \right) \right] d\tau' \tag{4.46}$$

and

$$\vec{A}(\vec{r}, t) = \frac{\mu_0}{4\pi} \int \left[\left(\frac{\vec{J}(r, t_r)}{R} \right) \right] d\tau' \tag{4.47}$$

The corresponding electric and magnetic potentials are given by

$$\vec{E} = -\vec{\nabla}V - \frac{\partial \vec{A}}{\partial t} \quad \text{and} \quad \vec{B} = \vec{\nabla} \times \vec{A}$$

Let us first evaluate

$$\vec{\nabla}V = \frac{1}{4\pi\epsilon_0} \int \left[\vec{\nabla} \left(\frac{\rho(r, t_r)}{R} \right) \right] d\tau'$$

For two scalar functions ψ and ϕ, we can write:

$$\vec{\nabla}(\psi\phi) = \phi\vec{\nabla}\psi + \psi\vec{\nabla}\phi$$

We have already derived

$$\vec{\nabla}V = \frac{1}{4\pi\epsilon_0} \int \left[\frac{-\dot{\rho}}{c} \frac{\hat{R}}{R} - \rho \frac{\hat{R}}{R^2} \right] d\tau'$$

$$\frac{\partial \vec{A}}{\partial t} = \frac{\mu_0}{4\pi} \int \frac{\frac{\partial \vec{J}}{\partial t}}{R} d\tau'$$

$$\frac{\partial \vec{A}}{\partial t} = \frac{\mu_0}{4\pi} \int \frac{\dot{\vec{J}}}{R} d\tau'$$

Using these equations in the electric field equation, we get

$$\vec{E} = \frac{-1}{4\pi\epsilon_0} \int \left[\frac{-\dot{\rho}}{c} \frac{\hat{R}}{R} - \rho \frac{\hat{R}}{R^2} \right] d\tau' - \frac{\mu_0}{4\pi} \int \frac{\dot{\vec{J}}}{R} d\tau'$$

or

$$\vec{E} = \frac{1}{4\pi\epsilon_0} \int \left\{ \left[\rho \frac{\hat{R}}{R^2} + \frac{\dot{\rho}}{c}\frac{\hat{R}}{R} \right] d\tau' - \int \frac{\dot{\vec{J}}}{c^2 R} d\tau' \right\} \tag{4.48}$$

which is the time-dependent generalization of the Coulomb's law.

Magnetic Field

The magnetic field $\vec{B}$ can be determined using the Biot-Savart law and its generalization to account for time-varying currents. To derive this, we begin with the identity:
$$\vec{B}$$

$$\vec{\nabla} \times \left(f\vec{T} \right) = f\left(\vec{\nabla} \times \vec{T} \right) - \vec{T} \times \vec{\nabla}f$$

Here,

$$f \rightarrow \frac{1}{R} \quad \text{and} \quad \vec{T} \rightarrow \vec{J}$$

$$\vec{\nabla} \times \vec{A} = \frac{\mu_0}{4\pi} \int \left[\vec{\nabla} \times \left(\frac{\vec{J}}{R} \right) \right] d\tau'$$

$$\vec{\nabla} \times \vec{A} = \frac{\mu_0}{4\pi} \int \left[\frac{1}{R}\left(\vec{\nabla} \times \vec{J} \right) - \vec{J} \times \vec{\nabla}\left(\frac{1}{R} \right) \right] d\tau'$$

$$\vec{\nabla} \times \vec{J} = \begin{vmatrix} \hat{i} & \hat{j} & \hat{k} \\ \frac{\partial}{\partial x} & \frac{\partial}{\partial y} & \frac{\partial}{\partial z} \\ J_x & J_y & J_z \end{vmatrix}$$

$$\left(\vec{\nabla} \times \vec{J} \right)_x = \frac{\partial J_z}{\partial y} - \frac{\partial J_y}{\partial z}$$

and we can write $\frac{\partial J_z}{\partial y}$ as:

$$\frac{\partial J_z}{\partial y} = \frac{\partial J_z}{\partial t_r}\frac{\partial t_r}{\partial y} = \dot{J}_z\frac{\partial t_r}{\partial y}$$

$$t_r = t - \frac{R}{c}$$

$$\frac{\partial J_z}{\partial y} = \frac{-1}{c}\dot{J}_z\frac{\partial R}{\partial y}$$

$$\left(\vec{\nabla} \times \vec{J} \right)_x = \frac{-1}{c}\left(\dot{J}_z\frac{\partial R}{\partial y} - \dot{J}_y\frac{\partial R}{\partial z} \right)$$

$$\left(\vec{\nabla} \times \vec{J} \right)_x = \frac{1}{c}\left(\dot{J}_y\frac{\partial R}{\partial z} - \dot{J}_z\frac{\partial R}{\partial y} \right)$$

and

$$\vec{J} \times \vec{\nabla} R = \begin{vmatrix} \hat{i} & \hat{j} & \hat{k} \\ \dot{J}_x & \dot{J}_y & \dot{J}_z \\ \frac{\partial R}{\partial x} & \frac{\partial R}{\partial y} & \frac{\partial R}{\partial z} \end{vmatrix}$$

$$\left(\vec{\nabla} \times \vec{J} \right)_x = \frac{1}{c} \left(\dot{\vec{J}} \times \left(\vec{\nabla} R \right) \right)_x$$

$$\vec{\nabla} R = \hat{R}$$

Similarly, we can prove for y and z components

$$\vec{\nabla} \times \vec{J} = \frac{1}{c} \dot{\vec{J}} \times \hat{R}$$

Also,

$$\vec{\nabla} \left(\frac{1}{R} \right) = \frac{-\hat{R}}{R^2}$$

$$\vec{B}(\vec{r}, t) = \frac{\mu_0}{4\pi} \int \left[\frac{\vec{J}}{R^2} + \frac{1}{cR} \dot{\vec{J}} \right] \times \hat{R} d\tau' \tag{4.49}$$

This is the generalization of the Biot-Savart law. This equation represents the magnetic field generated by a time-dependent current distribution and generalizes the Biot-Savart law to account for the effects of changing currents over time.

Example 4.10 Suppose the current density is constant in time, so $\rho(\vec{r}, t) = \rho(\vec{r}, 0) + \dot{\rho}(\vec{r}, 0)t$. Using the Jefimenko equation for the electric field, show that $\vec{E}(\vec{r}, t) = \frac{1}{4\pi\epsilon_0} \int \frac{\rho(\vec{r}, t_r)}{r^2} \hat{r} d\tau'$.

Solution:

We know that Jefimenko's equation for the electric field is given by

$$\vec{E}(\vec{r}, t) = \frac{1}{4\pi\epsilon_0} \int \left[\frac{\rho(\vec{r}', t_r)}{r^2} \hat{r} + \frac{\dot{\rho}(\vec{r}', t_r)}{cr} \hat{r} - \frac{\dot{\vec{J}}(\vec{r}', t_r)}{c^2 r} \right] d\tau'$$

In this case $\dot{\rho}(\vec{r}, t) = \dot{\rho}(\vec{r}, 0)$ and $\dot{\vec{J}}(\vec{r} \cdot t) = 0$

$$\vec{E}(\vec{r}, t) = \frac{1}{4\pi\epsilon_0} \int \left[\frac{\rho(\vec{r}', 0) + \dot{\rho}(\vec{r}', 0)t_r}{r^2} \hat{r} + \frac{\dot{\rho}(\vec{r}', 0)}{cr} \hat{r} - \frac{0}{c^2 r} \right] d\tau'$$

$$\vec{E}(\vec{r}, t) = \frac{1}{4\pi\epsilon_0} \int \left[\frac{\rho(\vec{r}', 0) + \dot{\rho}(\vec{r}', 0)(t - \frac{r}{c})}{r^2} \hat{r} + \frac{\dot{\rho}(\vec{r}', 0)}{cr} \right] \hat{r} d\tau'$$

$$\vec{E}(\vec{r}, t) = \frac{1}{4\pi \epsilon_0} \int \left[\frac{\rho(\vec{r}', 0) + \dot{\rho}(\vec{r}', 0)t}{r^2} - \frac{\dot{\rho}(\vec{r}', 0)\frac{r}{c}}{r^2} + \frac{\dot{\rho}(\vec{r}', 0)}{cr} \right] \hat{r} d\tau'$$

Therefore,

$$\vec{E}(\vec{r}, t) = \frac{1}{4\pi \epsilon_0} \int \frac{\rho(\vec{r}, t_r)}{r^2} \hat{r} d\tau'.$$

4.7 The Lineard–Wiechert Potentials

In this case, we are mulling to evaluate the potential due to a point charge q, that is moving on a specified trajectory. Let $\vec{W}(t_r)$ is the position of charge q at time t_r. We can determine the retarded time as

$$\left| \vec{r} - \vec{W}(t_r) \right| = c(t - t_r) \tag{4.50}$$

and

$$\vec{R} = \vec{r} - \vec{W}(t_r)$$

At any particular instant of time, only one point on the trajectory will be in communication with $\vec{r}$. We can show the same with the simple logic (Fig. 4.1).

Let us assume that there are two points with the retarded times t_1 and t_2. Therefore, we can write

$$R_1 = c(t - t_1) \quad \text{and} \quad R_2 = c(t - t_2)$$

Fig. 4.1 A point charge q traversing a certain trajectory

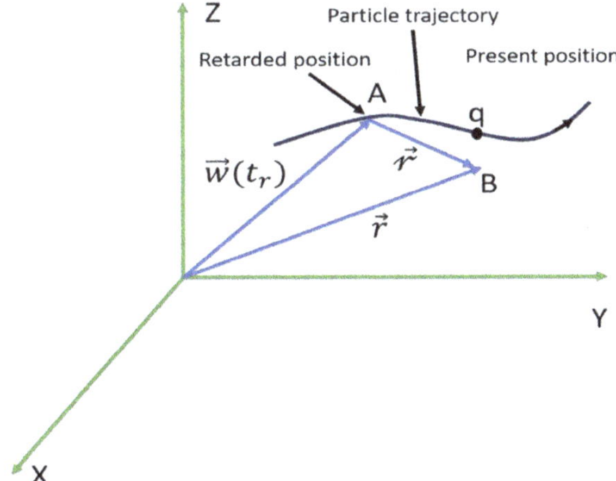

Hence, we write

$$R_1 - R_2 = c(t_2 - t_1) \tag{4.51}$$

It shows that the average velocity of the charge q in the direction of $\vec{r}$ is c, which violates special theory of relativity. The scalar retarded potential is given by

$$V(\vec{r}, t) = \frac{1}{4\pi \epsilon_0} \int \left[\left(\frac{\rho(\vec{r}', t_r)}{R} \right) \right] d\tau' \tag{4.52}$$

We cannot write the above equation as $V(\vec{r}, t) = \frac{1}{4\pi \epsilon_0} \frac{q}{R}$, because $\int [\rho(\vec{r}', t_r)] d\tau'$ is not equal to total charge of a particle, for total charge, $\rho(\vec{r}', t_r)$ we have to integrate over the entire distribution at one instant of time. But, $t_r = t - \frac{R}{c}$ will not allow us to evaluate ρ at different times for different parts of the configuration. However, if the source is moving, this will give a distorted picture of the total charge. For an extended particle, no matter how much small, the retardation throws a factor, $\vec{V}$ which is the velocity at a retarded time, because of motion as

$$\int [\rho(\vec{r}', t_r)] d\tau' = \frac{q}{1 - \hat{R} \cdot \frac{\vec{V}}{c}} \tag{4.53}$$

We will prove Eq. (4.53) as follows.

Proof This is the pure geometrical effect. A train approaching towards you looks a little longer than what really it is. In the interval chosen it takes light to travel the extra distance L', meanwhile the train itself moves a distance $L' - L$. (Fig. 4.2)

$$\frac{L'}{c} = \frac{L' - L}{V}$$

$$\frac{L'}{c} - \frac{L'}{V} = \frac{-L}{V}$$

$$L' \left(\frac{1}{c} - \frac{1}{V} \right) = -\frac{L}{V} \tag{4.54}$$

which implies,

$$L' = \frac{L}{1 - \frac{V}{c}} \tag{4.55}$$

So, approaching train appears longer by a factor of $\left(1 - \frac{V}{c}\right)^{-1}$. By constraint a train going away may, therefore look shorter by a factor $\left(1 + \frac{V}{c}\right)^{-1} \left(As \frac{L'}{c} = \frac{L-L'}{V} \right)$. For instance, if the train velocity makes an angle θ with the line of sight,

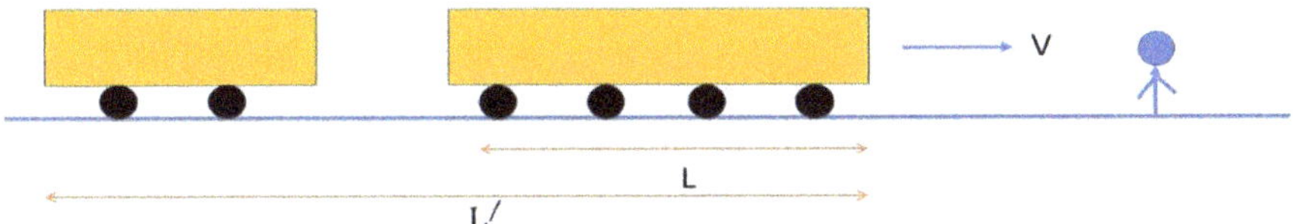

Fig. 4.2 Visualization of relativistic effects when an object approaches or recedes from an observer

$$\frac{L' \cos(\theta)}{C} = \frac{L' - L}{V} \tag{4.56}$$

$$L' = \frac{L}{1 - \frac{V \cos(\theta)}{c}} \tag{4.57}$$

The effect doesn't distort the dimensions normal to the direction of motion. The apparent volume of the train τ' related to the actual volume τ, is therefore

$$\tau' = \frac{\tau}{1 - \frac{\hat{R} \cdot \vec{V}}{c}} \tag{4.58}$$

Here, $\hat{R}$ is the unit vector from the train to the observer. Whenever, you evaluate an integral of the type $\int \left[\rho(\vec{r}', t_r) \right] d\tau'$, the effective volume is modified by the factor $\left(1 - \frac{\hat{R} \cdot \vec{V}}{c} \right)$. This factor has no reference with regard to size of the particle. Therefore, it follows that

$$V(\vec{r}, t) = \frac{1}{4\pi \epsilon_0} \frac{qc}{\left(Rc - \vec{R} \cdot \vec{V} \right)} \tag{4.59}$$

where $\vec{V}$ is the velocity at retarded time and $\vec{R}$ is the distance from retarded position to field point. Since, the current density of a rigid object is $\rho \vec{V}$,

$$\vec{A}(\vec{r}, t) = \frac{\mu_0}{4\pi} \int \left[\left(\frac{\rho(\vec{r}', t_r)}{R} \right) \vec{V}(t_r) \right] d\tau' \tag{4.60}$$

$$\vec{A}(\vec{r}, t) = \frac{\mu_0}{4\pi} \frac{\vec{V}}{R} \int \rho\left(\vec{r'}, t_r\right) d\tau'$$

$$\vec{A}(\vec{r}, t) = \frac{\mu_0}{4\pi} \frac{\vec{V}}{R} \frac{q}{1 - \frac{\hat{R}\cdot\vec{V}}{c}}$$

$$\vec{A}(\vec{r}, t) = \frac{\mu_0}{4\pi} \frac{q\vec{V}c}{\left(Rc - \vec{R}\cdot\vec{V}\right)} \tag{4.61}$$

$$\vec{A}(\vec{r}, t) = \mu_0\epsilon_0\vec{V}\left(\frac{1}{4\pi\epsilon_0} \frac{qc}{\left(Rc - \vec{R}\cdot\vec{V}\right)}\right)$$

$$\vec{A}(\vec{r}, t) = \frac{\vec{V}}{c^2} V(\vec{r}, t)$$

Here, $V(\vec{r}, t)$ and $\vec{A}(\vec{r}, t)$ are called as the Lineard–Wiechert potentials for a moving point charge.

4.8 The Fields of a Moving Point Charge

Fields $\vec{E}$ and $\vec{B}$ of a moving charge in arbitrary direction of motion are obtained by utilizing the following equations (Fig. 4.3).

$$\vec{E} = -\vec{\nabla}V - \frac{\partial\vec{A}}{\partial t} \quad \text{and} \quad \vec{B} = \vec{\nabla} \times \vec{A}$$

Here,

$$V(\vec{r}, t) = \frac{1}{4\pi\epsilon_0} \frac{qc}{\left(Rc - \vec{R}\cdot\vec{V}\right)}$$

$$\vec{A}(\vec{r}, t) = \frac{\vec{V}}{c^2} V(\vec{r}, t)$$

$\vec{R} = \vec{r} - \vec{W}(t_r)$ and $\vec{V} = \dot{\vec{W}}(t_r)$, both evaluated at retarded time and t_r itself depends on

$$\left|\vec{r} - \vec{W}(t_r)\right| = c(t - t_r)$$

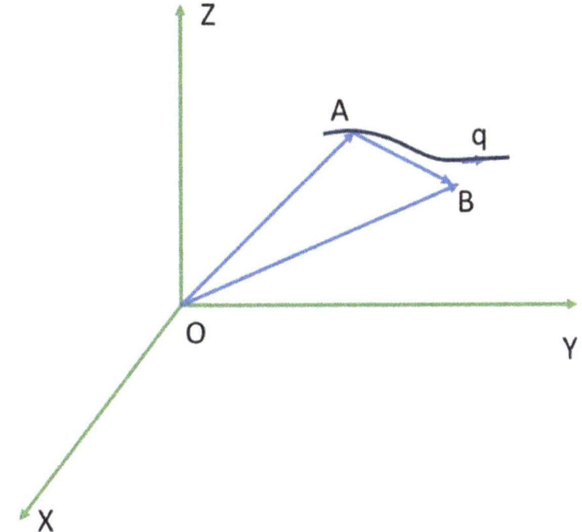

Fig. 4.3 A point charge q moving along a specified trajectory

Let us begin:

$$\vec{\nabla}V = \frac{-qc}{4\pi\,\epsilon_0}\,\frac{1}{\left(Rc - \vec{R}\cdot\vec{V}\right)^2}\,\vec{\nabla}\left(Rc - \vec{R}\cdot\vec{V}\right) \qquad (4.62)$$

$$\vec{\nabla}V = \frac{-c\vec{\nabla}R + \vec{\nabla}\vec{R}\cdot\vec{V}}{4\pi\,\epsilon_0\left(Rc - \vec{R}\cdot\vec{V}\right)^2}\,(qc) \qquad (4.63)$$

1st term of (4.63) can be written as:

$$\vec{\nabla}R = \vec{\nabla}c(t - t_r) = -c\vec{\nabla}(t_r) \qquad (4.64)$$

2nd term of (4.63) can be written as:

$$\vec{\nabla}\left(\vec{R}\cdot\vec{V}\right) = \left(\vec{R}\cdot\vec{\nabla}\right)\vec{V} + \left(\vec{V}\cdot\vec{\nabla}\right)\vec{R} + \vec{R}\times\left(\vec{\nabla}\times\vec{V}\right) + \vec{V}\times\left(\vec{\nabla}\times R\right) \qquad (4.65)$$

The 2nd term in turn contains the four terms. We will evaluate its terms one by one.

Term I:

$$\left(\vec{R}\cdot\vec{\nabla}\right)\vec{V} = \left(R_x\frac{\partial}{\partial x} + R_y\frac{\partial}{\partial y} + R_z\frac{\partial}{\partial z}\right)\vec{V}(t_r)$$

$$\left(\vec{R}\cdot\vec{\nabla}\right)\vec{V} = R_x\frac{d\vec{V}}{dt_r}\frac{\partial t_r}{\partial x} + R_y\frac{d\vec{V}}{dt_r}\frac{\partial t_r}{\partial y} + R_z\frac{d\vec{V}}{dt_r}\frac{\partial t_r}{\partial z}$$

$$= \frac{d\vec{V}}{dt_r}\left(R_x\frac{\partial t_r}{\partial x} + R_y\frac{\partial t_r}{\partial y} + R_z\frac{\partial t_r}{\partial z}\right)$$

$$\left(\vec{R}\cdot\vec{\nabla}\right)\vec{V} = \vec{a}\left(\vec{R}\cdot\vec{\nabla}(t_r)\right) \tag{4.66}$$

where $\vec{a} = \dot{\vec{V}}$; acceleration at retarded time.

Term II:

$$\left(\vec{V}\cdot\vec{\nabla}\right)\vec{R} = \left(\vec{V}\cdot\vec{\nabla}\right)\vec{r} - \left(\vec{V}\cdot\vec{\nabla}\right)\vec{W}$$

$$\left(\vec{V}\cdot\vec{\nabla}\right)\vec{r} = \left(V_x\frac{\partial}{\partial x} + V_y\frac{\partial}{\partial y} + V_z\frac{\partial}{\partial z}\right)(x\hat{i} + y\hat{j} + x\hat{z})$$

$$\left(\vec{V}\cdot\vec{\nabla}\right)\vec{r} = (V_x\hat{i} + V_y\hat{j} + V_z\hat{z}) = \vec{V}$$

$$\left(\vec{V}\cdot\vec{\nabla}\right)\vec{W}(t_r) = \left(V_x\frac{\partial}{\partial x} + V_y\frac{\partial}{\partial y} + V_z\frac{\partial}{\partial z}\right)\vec{W}(t_r)$$

$$\left(\vec{V}\cdot\vec{\nabla}\right)\vec{W}(t_r) = V_x\frac{d\vec{W}}{dt_r}\frac{\partial t_r}{\partial x} + V_y\frac{d\vec{W}}{dt_r}\frac{\partial t_r}{\partial y} + V_z\frac{d\vec{W}}{dt_r}\frac{\partial t_r}{\partial z}$$

$$\left(\vec{V}\cdot\vec{\nabla}\right)\vec{W}(t_r) = \vec{V}\left(\vec{V}\cdot\vec{\nabla}(t_r)\right) \tag{4.67}$$

Term III:

$$\vec{\nabla}\times\vec{V} = \left(\frac{\partial V_z}{\partial y} - \frac{\partial V_y}{\partial z}\right)\hat{i} + \left(\frac{\partial V_x}{\partial z} - \frac{\partial V_z}{\partial x}\right)\hat{j} + \left(\frac{\partial V_y}{\partial x} - \frac{\partial V_x}{\partial y}\right)\hat{k}$$

$$= \left(\frac{dV_z}{dt_r}\frac{\partial t_r}{\partial y} - \frac{dV_y}{dt_r}\frac{\partial t_r}{\partial z}\right)\hat{i} + \left(\frac{dV_x}{dt_r}\frac{\partial t_r}{\partial z} - \frac{dV_z}{dt_r}\frac{\partial t_r}{\partial x}\right)\hat{j}$$

$$+ \left(\frac{dV_y}{dt_r}\frac{\partial t_r}{\partial x} - \frac{dV_x}{dt_r}\frac{\partial t_r}{\partial y}\right)\hat{k}$$

$$= -\left[\left(a_y\frac{\partial t_r}{\partial z} - a_z\frac{\partial t_r}{\partial y}\right)\hat{i} - \left(a_x\frac{\partial t_r}{\partial z} - a_z\frac{\partial t_r}{\partial x}\right)\hat{j} + \left(a_x\frac{\partial t_r}{\partial y} - a_y\frac{\partial t_r}{\partial x}\right)\hat{k}\right]$$

$$\vec{\nabla}\times\vec{V} = -\vec{a}\times\vec{\nabla}(t_r) \tag{4.68}$$

Term IV:

$$\vec{\nabla}\times\vec{R} = \vec{\nabla}\times\left(\vec{r} - \vec{W}(t_r)\right)$$

$$\vec{\nabla}\times\vec{R} = \vec{\nabla}\times\vec{r} - \vec{\nabla}\times\vec{W}(t_r)$$

$$\vec{\nabla}\times\vec{R} = 0 - \vec{\nabla}\times\vec{W}$$

We can solve $\vec{\nabla} \times \vec{W}$ in the same way as we did for $\vec{\nabla} \times \vec{V}$ for term III. Therefore, we can write:

$$\vec{\nabla} \times \vec{W} = -\vec{V} \times \vec{\nabla}(t_{\mathrm{r}})$$

which implies

$$\vec{\nabla} \times \vec{R} = \vec{V} \times \vec{\nabla}(t_{\mathrm{r}}) \tag{4.69}$$

Substitute Eqs. (4.66), (4.67), (4.68) and (4.69) in Eq. (4.65), we obtain

$$\vec{\nabla}\left(\vec{R} \cdot \vec{V}\right) = \vec{a}\left(\vec{R} \cdot \vec{\nabla}(t_{\mathrm{r}})\right) + \vec{V} - \vec{V}\left(\vec{V} \cdot \vec{\nabla}(t_{\mathrm{r}})\right)$$
$$- \vec{R} \times \left(\vec{a} \times \vec{\nabla}(t_{\mathrm{r}})\right) + \vec{V} \times \left(\vec{V} \times \vec{\nabla}(t_{\mathrm{r}})\right)$$

Using the identity

$$\vec{A} \times \left(\vec{B} \times \vec{C}\right) = \vec{B}\left(\vec{A} \cdot \vec{C}\right) - \vec{C}\left(\vec{A} \cdot \vec{B}\right)$$
$$\vec{R} \times \left(\vec{a} \times \vec{\nabla}(t_{\mathrm{r}})\right) = \vec{a}\left(\vec{R} \cdot \vec{\nabla}(t_{\mathrm{r}})\right) - \vec{\nabla}(t_{\mathrm{r}})\left(\vec{R} \cdot \vec{a}\right)$$

and

$$\vec{V} \times \left(\vec{V} \times \vec{\nabla}(t_{\mathrm{r}})\right) = \vec{V}\left(\vec{V} \cdot \vec{\nabla}(t_{\mathrm{r}})\right) - \vec{\nabla}(t_{\mathrm{r}})\left(\vec{V} \cdot \vec{V}\right)$$

Putting these, relations

$$\vec{\nabla}\left(\vec{R} \cdot \vec{V}\right) = \vec{a}\left(\vec{R} \cdot \vec{\nabla}(t_{\mathrm{r}})\right) + \vec{V} - \vec{V}\left(\vec{V} \cdot \vec{\nabla}(t_{\mathrm{r}})\right) - \vec{a}\left(\vec{R} \cdot \vec{\nabla}(t_{\mathrm{r}})\right)$$
$$+ \vec{\nabla}(t_{\mathrm{r}})\left(\vec{R} \cdot \vec{a}\right) + \vec{V}\left(\vec{V} \cdot \vec{\nabla}(t_{\mathrm{r}})\right) - \vec{\nabla}(t_{\mathrm{r}})V^2$$
$$\vec{\nabla}\left(\vec{R} \cdot \vec{V}\right) = \vec{V} + \left(\vec{R} \cdot \vec{a} - V^2\right)\vec{\nabla}(t_{\mathrm{r}}) \tag{4.70}$$

Substitute Eqs. (4.64) and (4.70) in Eq. (4.63), we get

$$\vec{\nabla}V = \frac{qc}{4\pi\epsilon_0} \frac{1}{\left(Rc - \vec{R} \cdot \vec{V}\right)^2}\left[\vec{V} + \left(c^2 - V^2 + \vec{R} \cdot \vec{a}\right)\vec{\nabla}(t_{\mathrm{r}})\right] \tag{4.71}$$

Now, we have to evaluate $\vec{\nabla}(t_{\mathrm{r}})$

$$-c\vec{\nabla}(t_{\mathrm{r}}) = \vec{\nabla}R = \vec{\nabla}\sqrt{\vec{R} \cdot \vec{R}} = \frac{1}{2\sqrt{\vec{R} \cdot \vec{R}}}\vec{\nabla}\left(\vec{R} \cdot \vec{R}\right)$$

$$-c\vec{V}(t_{\mathrm{r}}) = \frac{1}{2R}\left\{\vec{R}\times\left(\vec{\nabla}\times\vec{R}\right) + \vec{R}\times\left(\vec{\nabla}\times\vec{R}\right) + \left(\vec{R}\cdot\vec{\nabla}\right)\vec{R} + \left(\vec{R}\cdot\vec{\nabla}\right)\vec{R}\right\}$$

$$-c\vec{V}(t_{\mathrm{r}}) = \frac{1}{R}\left\{\left(\vec{R}\cdot\vec{\nabla}\right)\vec{R} + \vec{R}\times\left(\vec{\nabla}\times\vec{R}\right)\right\}$$

From the above equation, the term $\left(\vec{R}\cdot\vec{\nabla}\right)\vec{R}$ can be written as:

$$\left(\vec{R}\cdot\vec{\nabla}\right)\vec{R} = \left(\vec{R}\cdot\vec{\nabla}\right)\vec{r} - \left(\vec{R}\cdot\vec{\nabla}\right)\vec{W}$$

$$\left(\vec{R}\cdot\vec{\nabla}\right)\vec{R} = \vec{R} - \vec{V}\left(\vec{R}\cdot\vec{\nabla}(t_{\mathrm{r}})\right)$$

We have already evaluated, $\vec{\nabla}\times\vec{R} = \vec{V}\times\vec{\nabla}t_{\mathrm{r}}, \vec{\nabla}\times\vec{R} = \vec{V}\times\vec{\nabla}t_{\mathrm{r}}$

Put all these expressions in $-c\vec{V}(t_{\mathrm{r}})$, we get

$$-c\vec{V}(t_{\mathrm{r}}) = \frac{1}{R}\left[\vec{R} - \vec{V}\left(\vec{R}\cdot\vec{\nabla}(t_{\mathrm{r}}) + \vec{R}\times\left(\vec{V}\times\vec{\nabla}t_{\mathrm{r}}\right)\right)\right]$$

$$\vec{A}\times\left(\vec{B}\times\vec{C}\right) = \vec{B}\left(\vec{A}\cdot\vec{C}\right) - \vec{C}\left(\vec{A}\cdot\vec{B}\right)$$

$$-c\vec{V}(t_{\mathrm{r}}) = \frac{1}{R}\left[\vec{R} - \vec{V}\left(\vec{R}\cdot\vec{\nabla}(t_{\mathrm{r}})\right) + \vec{V}\left(\vec{R}\cdot\vec{\nabla}t_{\mathrm{r}}\right) - \vec{\nabla}(t_{\mathrm{r}})\left(\vec{R}\cdot\vec{V}\right)\right]$$

$$-c\vec{V}(t_{\mathrm{r}}) = \frac{1}{R}\left[\vec{R} - \left(\vec{R}\cdot\vec{V}\right)\vec{\nabla}(t_{\mathrm{r}})\right]$$

$$-c\vec{V}(t_{\mathrm{r}}) = \frac{-\vec{R}}{Rc - \vec{R}\cdot\vec{V}} \tag{4.72}$$

Substitute Eq. (4.72) in Eq. (4.71), we have

$$\vec{\nabla}V = \frac{qc}{4\pi\epsilon_0}\frac{1}{\left(Rc - \vec{R}\cdot\vec{V}\right)^2}\left[\vec{V} + \left(c^2 - V^2 + \vec{R}\cdot\vec{a}\right)\left(\frac{-\vec{R}}{Rc - \vec{R}\cdot\vec{V}}\right)\right]$$

$$\vec{\nabla}V = \frac{qc}{4\pi\epsilon_0}\frac{1}{\left(Rc - \vec{R}\cdot\vec{V}\right)^3}\left[\left(Rc - \vec{R}\cdot\vec{V}\right)\vec{V} - \left(c^2 - V^2 + \vec{R}\cdot\vec{a}\right)\vec{R}\right] \tag{4.73}$$

Similarly, we can prove that

$$\frac{\partial\vec{A}}{\partial t} = \frac{1}{4\pi\epsilon_0}\frac{qc}{\left(Rc - \vec{R}\cdot\vec{V}\right)^3}$$

$$\times\left[\left(Rc - \vec{R}\cdot\vec{V}\right)\left(-\vec{V} + \frac{\vec{R}\cdot\vec{a}}{c}\right) + \frac{R}{c}\left(c^2 - V^2 + \vec{R}\cdot\vec{a}\right)\vec{V}\right] \tag{4.74}$$

Combining these results in $\vec{E} = -\vec{\nabla}V - \frac{\partial\vec{A}}{\partial t}$,

$$\vec{E} = \frac{-qc}{4\pi\epsilon_0} \frac{1}{\left(Rc - \vec{R}\cdot\vec{V}\right)^3}$$

$$\times \left[-\left(c^2 - V^2 + \vec{R}\cdot\vec{a}\right)\vec{R} + \left(Rc - \vec{R}\cdot\vec{V}\right)\frac{R\vec{a}}{c} + \frac{R}{c}\left(c^2 - V^2 + \vec{R}\cdot\vec{a}\right)\vec{V} \right]$$

$$\vec{E} = \frac{qc}{4\pi\epsilon_0} \frac{1}{\left(Rc - \vec{R}\cdot\vec{V}\right)^3}$$

$$\times \left[\left(c^2 - V^2\right)\left(\vec{R} - \frac{R}{c}\vec{V}\right) + \vec{R}\cdot\vec{a}\left(\vec{R} - \frac{R}{c}\vec{V}\right) - \left(Rc - \vec{R}.\vec{V}\right)\frac{R\vec{a}}{c} \right]$$

$$\vec{E} = \frac{qc}{4\pi\epsilon_0} \frac{1}{\left(Rc - \vec{R}\cdot\vec{V}\right)^3}$$

$$\times \left[\frac{R}{c}\left(c^2 - V^2\right)\left(c\hat{R} - \vec{V}\right) + \frac{R}{c}\vec{R}\cdot\vec{a}\left(c\hat{R} - \vec{V}\right) - \vec{R}\left(c\hat{R} - \vec{V}\right)\frac{R\vec{a}}{c} \right]$$

If $\vec{u} = c\hat{R} - \vec{V}\vec{u} = c\hat{R} - \vec{V}$

$$\vec{E} = \frac{qc}{4\pi\epsilon_0} \frac{\frac{R}{c}}{\left(Rc - \vec{R}\cdot\vec{V}\right)^3}\left[\left(c^2 - V^2\right)\vec{u} + \left(\vec{R}\cdot\vec{a}\right)\vec{u} - \left(\vec{R}\cdot\vec{u}\right)\vec{a} \right]$$

$$\vec{E} = \frac{q}{4\pi\epsilon_0} \frac{R}{\left(\vec{R}\cdot\vec{u}\right)^3}\left[\left(c^2 - V^2\right)\vec{u} + \vec{R} \times (\vec{u} \times \vec{a}) \right] \tag{4.75}$$

This is the electric field due to the moving point charge. The first term in $\vec{E}$ is the velocity-dependent field and is called as generalized Coulomb's field. The 2nd term is known as the radiation field (accelerated dependent field).

The magnetic field is given by:

$$\vec{B} = \vec{\nabla} \times \vec{A}$$

$$\vec{B} = \vec{\nabla} \times \left(\frac{\vec{V}}{c^2} V(\vec{r}, t) \right) = \frac{1}{c^2}\left[V\left(\vec{\nabla} \times \vec{V}\right) - \vec{V} \times \vec{\nabla}V \right]$$

We have proved above that

$$\vec{\nabla} \times \vec{V} = -\vec{a} \times \vec{\nabla}(t_r)$$

We also know that

$$\vec{\nabla}(t_r) = \frac{-\vec{R}}{Rc - \vec{R}\cdot\vec{V}}$$

Putting all these together in $\vec{\nabla} \times \vec{A}$, we get

$$\vec{\nabla} \times \vec{A} = \frac{-1}{c} \frac{q}{4\pi\epsilon_0} \frac{R}{(\vec{R} \cdot \vec{u})^3} \hat{R} \times \left[(c^2 - V^2)\vec{V} + (\vec{R} \cdot \vec{a})\vec{V} + (\vec{R} \cdot \vec{u})\vec{a} \right] \quad (4.76)$$

Further, we have $\vec{u} = c\hat{R} - \vec{V}$, implies $\vec{V} = c\hat{R} - \vec{u}$. Again, making use of

$$\vec{A} \times (\vec{B} \times \vec{C}) = \vec{B}(\vec{A} \cdot \vec{C}) - \vec{C}(\vec{A} \cdot \vec{B})$$

$$\vec{B}(\vec{r}, t) = \frac{1}{c}\hat{R} \times \vec{E}(\vec{r}, t) \quad (4.77)$$

As we are conversant with the fact that the magnetic field due to a point charge is always perpendicular to the direction of electric field and to the vector from the retarded field, $\vec{E}(\vec{r}, t)$. In case $\vec{V}$ and $\vec{a}$ both are zero, therefore, we can write from Eq. (4.75)

$$\vec{E} = \frac{q}{4\pi\epsilon_0} \frac{1}{R^2}\hat{R} \quad (4.78)$$

The second term of $\vec{E}$ is inversely proportional to R, therefore, it is the dominant term at large distance. Similarly, we can calculate force on test charged q due to charge q in electrodynamics and is hence given by

$$\vec{F} = q\left(\vec{E} + \vec{V} \times \vec{B}\right) \quad (4.79)$$

which can be written as follows

$$\vec{F} = \frac{qQ}{4\pi\epsilon_0} \frac{R}{\left(\vec{R} \cdot \vec{u}\right)^3}$$

$$\times \left[(c^2 - V^2)\vec{u} + \vec{R} \times (\vec{u} \times \vec{a}) + \frac{\vec{V}}{c} \times \left[\hat{R} \times \{(c^2 - V^2)\vec{u}\} + \vec{R} \times (\vec{u} \times \vec{a}) \right] \right]$$

$$(4.80)$$

Example 4.11 Suppose you take a plastic ring of radius a and glue charge on it, so that the line charge density is $\lambda_0 |sin(\theta/2)|$. Then you spin the loop about its axis at an angular velocity ω. Find the vector potentials at the center of the ring.

Solution:

We have a plastic ring of radius a with a line charge density given by:

$$\lambda = \lambda_0 |sin(\theta/2)|.$$

The vector potential at the center is given by:

$$\vec{A} = \frac{\mu_0}{4\pi} \int \frac{\lambda \vec{v} d\vec{l}}{r}$$

Since the ring is rotating at $\vec{\omega}$, charge element at θ has velocity:

$$\vec{v} = \omega a(-sin\theta \hat{x} + cos\theta \hat{y})$$

$$\int_0^{2\pi} sin(\theta/2)sin(\theta + \omega t_r)d\theta = \frac{1}{2}\int_0^{2\pi}\left[cos\left(\frac{\theta}{2} + \omega t_r\right) - cos\left(\frac{3\theta}{2} + \omega t_r\right)\right]d\theta$$

$$= \frac{1}{2}\left[2sin\left(\frac{\theta}{2} + \omega t_r\right) - \frac{2}{3}sin\left(\frac{3\theta}{2} + \omega t_r\right)\right]\Big|_0^{2\pi}$$

$$= sin(\pi + \omega t_r) - sin(\omega t_r) - \frac{1}{3}sin(3\pi + \omega t_r) + \frac{1}{3}sin(\omega t_r)$$

$$= -2sin(\omega t_r) + \frac{2}{3}sin(\omega t_r) = -\frac{4}{3}sin(\omega t_r)$$

$$\int_0^{2\pi} sin(\theta/2)sin(\theta + \omega t_r)d\theta = \frac{1}{2}\int_0^{2\pi}\left[-sin\left(\frac{\theta}{2} + \omega t_r\right) + sin\left(\frac{3\theta}{2} + \omega t_r\right)\right]d\theta$$

$$\frac{1}{2}\left[2cos\left(\frac{\theta}{2} + \omega t_r\right) - \frac{2}{3}cos\left(\frac{3\theta}{2} + \omega t_r\right)\right]\Big|_0^{2\pi}$$

$$= cos(\pi + \omega t_r) - cos(\omega t_r) - \frac{1}{3}cos(3\pi + \omega t_r) + \frac{1}{3}cos(\omega t_r)$$

$$= -2cos(\omega t_r) + \frac{2}{3}cos(\omega t_r) = -\frac{4}{3}cos(\omega t_r)$$

$$\vec{A}(t) = \frac{\mu_0 \lambda_0 \omega a}{4\pi}\left(\frac{4}{3}[sin(\omega t_r)\hat{x} - cos(\omega t_r)\hat{y}]\right)$$

$$\vec{A}(t) = \frac{\mu_0 \lambda_0 \omega a}{3\pi}\left\{sin\left[\omega(t - \frac{a}{c})\hat{x} - cos\left[\omega\left(t - \frac{a}{c}\right)\hat{y}\right]\right]\right\}$$

Example 4.12 A particle of charge q moves in a circle of radius a at constant angular velocity ω. (Assume that the circle lies in the xy plane, centered at the origin, and at time $(t = 0)$ the charge is at $(a, 0)$, on the positive x-axis). Find the Lienard-Wiechert potential for points on the z-axis.

Solution:

At a time t the charge is at $\vec{r}(t) = a[cos(\omega t)\hat{x} + sin(\omega t)\hat{y}]$ and we can write

$$\vec{v}(t) = \omega a[-sin(\omega t)\hat{x} + cos(\omega t)\hat{y}]$$

Hence,

$$\vec{r} = z\hat{z} - a[\cos(\omega t_r)\hat{x} + \sin(\omega t_r)\hat{y}]$$

$$r^2 = z^2 + a^2, \text{ and } r = \sqrt{z^2 + a^2}$$

$$\hat{r} \cdot \vec{v} = \frac{1}{r}(\hat{r} \cdot \vec{v})\frac{1}{r}\{-\omega a^2[-\sin(\omega t_r)\cos(\omega t_r) + \sin(\omega t_r)\cos(\omega t_r)]\} = 0,$$

so,

$$\left(1 - \frac{\hat{r} \cdot \vec{v}}{c}\right) = 1$$

Therefore,

$$V(z, t) = \frac{1}{4\pi\epsilon_0}\frac{q}{\sqrt{z^2 + a^2}}, \vec{A}(z, t) = \frac{q\omega a}{4\pi\epsilon_0 c^2\sqrt{z^2 + a^2}}[-\sin(\omega t_r)\hat{x} + \cos(\omega t_r)\hat{y}].$$

Where,

$$t_r = t - \frac{\sqrt{z^2 + a^2}}{c}.$$

Example 4.13 A point charge q with velocity $\vec{V} = V\hat{i}$ and acceleration $\vec{a} = a\hat{i}$. A test charge Q is placed at $\vec{r} = x\hat{i} + y\hat{j} + z\hat{k}$. Find the force on the charge Q.

Solution:

The electric field due to a moving charge is

$$\vec{E} = \frac{qc}{4\pi\epsilon_0}\frac{\frac{R}{c}}{\left(Rc - \vec{R} \cdot \vec{V}\right)^3}\left[(c^2 - V^2)\vec{u} + \left(\vec{R} \cdot \vec{a}\right)\vec{u} - \left(\vec{R} \cdot \vec{u}\right)\vec{a}\right]$$

$\vec{R} = \vec{r} - \vec{W}(t_r)$ is the displacement vector from the retarded position $\vec{W}(t_r)$ of the moving charge to the observation point $\vec{r}$.

$\vec{u} = c\hat{R} - \vec{V}$ with $\hat{R} = \frac{\vec{R}}{R}$.

The magnetic field is given by $\vec{W}(t_r) = Vt_r\hat{i}$

$$\vec{R} = (x - Vt_r)\hat{i} + y\hat{j} + z\hat{k}$$

$$R = \sqrt{(x - Vt_r)^2 + y^2 + z^2}$$

The unit vector $\hat{R}$ becomes

$$\hat{R} = \frac{(x - Vt_{\mathrm{r}})\hat{i} + y\hat{j} + z\hat{k}}{\sqrt{(x - Vt_{\mathrm{r}})^2 + y^2 + z^2}}$$

The unit vector $\vec{u} = c\hat{R} - V\hat{i}$. The magnetic due to the moving charge is given by

$$\vec{B}(\vec{r}, t) = \frac{1}{c}\hat{R} \times \vec{E}(\vec{r}, t)$$

$$\vec{F} = Q\left(\vec{E} + \vec{V} \times \vec{B}\right)$$

$$\vec{F} = \left(\frac{qQ}{4\pi\epsilon_0} \frac{R}{\left(R - \vec{R} \cdot \vec{V}/c^2\right)^3}\left[(c^2 - V^2)\vec{u} + \vec{u} \times (\vec{u} \times \vec{a})\right]\right)$$

Simplifying this force requires evaluating $\vec{u}$ and the terms $(c^2 - V^2)\vec{u}$ and $\vec{u} \times (\vec{u} \times \vec{a})$ explicitly, along with the identity.

$$\vec{u} \times (\vec{u} \times \vec{a}) = (\vec{u} \cdot \vec{a})\vec{u} - \vec{a}(\vec{u} \cdot \vec{u})$$

where $\vec{u} \cdot \vec{u} = c^2 - V^2$. Therefore, the expression for the force $\vec{F}$ on the test charge Q due to a moving charge q is:

$$\vec{F} = \left(\frac{qQ}{4\pi\epsilon_0} \frac{1}{\left(R - \vec{R} \cdot \vec{V}/c^2\right)^3}\left[(c^2 - V^2)\vec{u} + (\vec{u} \cdot \vec{a})\vec{u} - \vec{a}(c^2 - V^2)\right]\right)$$

Unsolved Problems

Problem 4.1 An atom of atomic number Z can be modelled as a point charge surrounded by a rigid uniformly negatively charged solid sphere of radius R. The electric polarizability α of this system is defined as $\alpha = \frac{P_E}{E}$. Where P_E is the dipole moment induced on application of electric field E which is small compared to the binding electric field inside the atom. Calculate the value of α.

Ans. $4\pi\epsilon_0 R^3$.

Problem 4.2 A scalar potential $V = V_0 e^{-\beta r}$ and vector potential $\vec{A} = A_0 e^{-\beta r}\hat{r}$ are defined for a region of space, where $\beta = 1 \text{ m}^{-1}$.

(a) Check if the Lorentz gauge condition $\vec{\nabla} \cdot \vec{A} + \frac{1}{c^2}\frac{\partial V}{\partial t} = 0$ holds.
(b) If not apply a gauge transformation to make V and $\vec{A}$ satisfy the Lorentz gauge.

Problem 4.3 Consider a cylindrical shell of radius R and height h centred along the z-axis carrying a uniform density $\vec{K} = K\hat{\phi}$. Calculate the vector potential $\vec{A}$ at a point on the z-axis at a distance z_0 from the centre of the shell.

Ans. $\frac{\mu_0 KRh}{2\sqrt{z_0^2 + R^2}}\hat{z}$.

Problem 4.4 Consider a static potential configuration where $V(\vec{r}) = \frac{q}{4\pi\epsilon_0 r}$ and $\vec{A} = \vec{A}_0$ a constant vector potential. Determine a gauge transformation that changes $\vec{A}$ to zero while leaving the electric and magnetic fields unchanged.

   Ans. $\vec{A}' = 0, V' = 0$.

Problem 4.5 Consider a circular loop of radius a carrying an oscillating current $I(r') = I_0\cos(\omega t')$. The loop lies in the xy-plane centred at the origin. Find the retarded potential $\vec{A}(\vec{r}, t)$ at a point on the z-axis a distance z from the loop.

   Ans. $\frac{\mu_0 I_0 a}{4\pi\sqrt{z^2+R^2}}\cos\left(\omega\left(t - \frac{\sqrt{z^2+R^2}}{c}\right)\right)\hat{\phi}$.

Problem 4.6 A point charge q moves with a constant velocity $\vec{v}$ along the x-axis. The position of the charge at any time is $W_x(t) = vt$. Calculate the electric field $\vec{E}(\vec{r}, t)$ at an observation point $\vec{r} = \left(x\hat{i} + y\hat{j} + z\hat{k}\right)$. Using the Lienard-Wiechert formula for the electric field of a moving charge.

   Ans. $\vec{E} = \frac{q}{4\pi\epsilon_0}\frac{R}{(Rc - \vec{R}\cdot\vec{V})^3}\left[\left(c^2 - V^2\right)\left(c\hat{R} - \vec{v}\right)\right]$.

Problem 4.7 For a continuous current density $\vec{J}(\vec{r}')$, derive the multipole expansion for the vector potential $\vec{A}(\vec{r})$ at a point $\vec{r}$ far from the current distribution.

   Ans. $\vec{A}(\vec{r}) = \frac{\mu_0}{4\pi}\int\frac{\vec{J}(\vec{r}')}{|\vec{r}-\vec{r}'|}d^3\vec{r}'$.

Problem 4.8 Show that the Coulomb gauge condition $\vec{\nabla}\cdot\vec{A} = 0$ does not uniquely determine $\vec{A}$.

Problem 4.9 Using the retarded potentials, derive the expressions for the scaler potential V and the vector potential $\vec{A}$ due to a continuous charge density $\rho(\vec{r}', t)$ and the current density $\vec{J}(\vec{r}')$.

Problem 4.10 Derive the expressions for the scalar and vector potentials in the Coulomb's gauge for a time-dependent charge distribution.

4.9   Summary

- **Introduction to Magnetic Phenomena**: We explored the transition from electrostatic to magnetostatic fields, emphasizing that magnetic fields arise from currents, contrasting with electric fields from charges. Potential formulations simplify theoretical understanding, but magnetic fields are more practical for measurement.
- **Vector Potential**: Defined magnetic fields using the vector potential $\vec{A}$ through the Biot-Savart law, which enables magnetic field calculation from current density. This approach parallels electrostatic potential V, aiding theoretical analysis and simplifying calculations.

- **Potentials and Fields Formulations in Electrodynamics**: Maxwell's equations extend to dynamic fields, linking electric and magnetic fields with time variations. Notably, electric and magnetic fields are redefined: $\vec{E} = -\vec{\nabla}V - \frac{\partial \vec{A}}{\partial t}$ and $\vec{B} = \vec{\nabla} \times \vec{A}$.
- **Gauge Transformations**: Introduced gauge freedom in electrodynamics, allowing modifications of V and $\vec{A}$ while preserving physical fields $\vec{E}$ and $\vec{B}$. Key gauges are the Coulomb gauge (for static problems) and Lorentz gauge (useful in relativistic contexts).
- **Coulomb's Gauge**: Focussed on simplifying scalar potential V in static cases, yielding Poisson's equation, a foundational equation in electrostatics for static charge distributions, while vector potential $\vec{A}$ requires more complex calculations.
- **Lorentz Gauge**: Treats V and $\vec{A}$ symmetrically, suitable for special relativity and representing potentials in inhomogeneous wave equations. The d'Alembertian operator unifies the theory and simplifies solutions in four-dimensional spacetime.
- **Continuous Charge Distributions and Retarded Potentials**: Retarded potentials are introduced for time-varying sources, accounting for the finite speed of electromagnetic propagation. These potentials adjust to dynamic sources, generalizing Coulomb's and Biot-Savart laws.
- **Jefimenko's Equations**: Derived solutions for electric and magnetic fields from time-dependent charges and currents, extending static field laws to dynamic contexts and enabling the calculation of fields influenced by source time variations.
- **Lienard-Wiechert Potentials**: Calculated fields for moving point charges, which factor in both velocity and acceleration, giving insight into radiated fields at large distances and supporting antenna theory.
- **Fields of a Moving Point Charge**: Developed expressions for electric and magnetic fields due to a point charge in motion. The electric field includes a velocity-dependent Coulomb component and an acceleration-dependent radiation component, while the magnetic field aligns perpendicularly to the electric field.
- Each topic builds a foundation in understanding magnetic fields through vector potentials, gauge choices and practical applications, with examples illustrating key concepts and calculations in electrodynamics.

Chapter 5
Relativistic Electrodynamics

Abstract In this chapter tensors extend the concept of vectors and serve as a foundational framework for analyzing physical phenomena across classical and relativistic contexts. Central to relativistic physics are Lorentz-invariant quantities, such as spacetime intervals, energy–momentum relations and Maxwell's equations, ensuring consistency across inertial frames. The four-dimensional dot product, utilizing the metric tensor $g_{\mu\nu}$, generalizes scalar products to 4D spacetime, underpinning key relations like $E^2 = p^2c^2 + m^2c^4$, which link energy, momentum and mass. Relativistic dynamics are governed by entities like four-velocity, four-acceleration, four-momentum and four-force, which integrate classical mechanics with relativistic corrections. Transformations of electric and magnetic fields between inertial frames illustrate the consistency of physical laws, while phenomena such as Lorentz contraction and time dilation underscore relativistic effects at high velocities. Lorentz invariance emerges as a unifying principle, maintaining the constancy of spacetime intervals, field relationships and invariant magnitudes of velocity and momentum four-vectors. The Lorentz gauge and continuity equation provide elegant representations of charge conservation and potential dynamics in spacetime. Relativistic kinetic energy relations integrate mass-energy equivalence, connecting classical energy concepts with relativistic formulations. This chapter bridges the gap between classical mechanics and special relativity, offering mathematical tools and insights for studying relativistic electrodynamic systems. It emphasizes the geometric and physical coherence of relativistic transformations, essential for understanding high-speed phenomena and electromagnetic field interactions.

Keywords Lorentz-invariant quantities · Metric tensor · Relativistic electrodynamic systems

R. Ahmad Parra et al., *A Systematic Approach to Electrodynamics*,
https://doi.org/10.1007/978-981-96-3408-8_5

5.1 Introduction

The incorporation of relativistic corrections into electrodynamics is essential because classical formulations fail to accurately describe electromagnetic phenomena at velocities approaching the speed of light. Without these corrections, key predictions about particle behaviour, field transformations, and radiation effects become increasingly inaccurate as speeds rise. Relativity provides the necessary framework to reconcile electromagnetism with the principles of special relativity, ensuring consistency across all inertial reference frames. This unification reveals how electric and magnetic fields fundamentally transform into one another when observed from different moving perspectives. To properly incorporate relativistic corrections into electrodynamics, we must formulate the theory using tensors.

Tensors can be dubbed as logical and natural generalization of vectors. The use of vectors is most profound in the mathematical study of wide range of physical phenomena. In a similar fashion, tensor analysis is widely applicable to various branches of physics. These applications can be broadly divided in to two main categories, viz., applications in non-relativistic physics and applications in the theories of relativity. In this chapter, we will provide a general definition of tensors followed by the algebra of tensors. Further, we will provide a brief description of the transformation laws of tensors.

5.2 Lorentz-Invariant Quantities

As discussed in Chap. 1, a vector is generally defined as follows:

$$\vec{a} = a_1\hat{x} + a_2\hat{y} + a_3\hat{z} \tag{5.1}$$

The dot product of above vector with itself is, therefore, written as under

$$\vec{a} \cdot \vec{a} = a_1^2 + a_2^2 + a_3^2 \tag{5.2}$$

It is a scalar quantity. However, if we have two different vectors $\vec{a}$ and $\vec{b}$, we can define the dot product as follows:

$$\vec{a} \cdot \vec{b} = a_1 b_1 + a_2 b_2 + a_3 b_3 \tag{5.3}$$

Or, in more compact notation we can illustrate above result as follows:

$$a_i \cdot b^i = a_1 b^1 + a_2 b^2 + a_3 b^3 \tag{5.4}$$

where a_i is the covariant vector and b^i is the contravariant vector.

In case of Euclidean 3D space there is no distinction between these two vectors. Therefore, we can write

$$a_i = (a_1, a_2, a_3)$$
$$a^i = (a_1, a_2, a_3)$$
$$(5.5)$$

Hence, Eq. (5.2) can be written as follows:

$$\vec{a} \cdot \vec{a} = a_i a^i = a_1 a^1 + a_2 a^2 + a_3 a^3 = a_1^2 + a_2^2 + a_3^2 \tag{5.6}$$

Of course, there is a plus sign between the components.
In Euclidian space, we have

$$r^2 = x^2 + y^2 + z^2 \tag{5.7}$$

which can be written as under

$$r^2 = x_1^2 + x_2^2 + x_3^2 \tag{5.8}$$

which resembles the dot product of a vector with itself. Thus, we can write it as under

$$r^2 = x_i x^i = x_1 x^1 + x_2 x^2 + x_3 x^3 \tag{5.9}$$

where range of i is equal to the dimensionality of space.
In Lorentz transformation, we know a quantity is Lorentz invariant when

$$(ct)^2 - r^2 = 0 \tag{5.10}$$

Since 0 is scalar, therefore, its value remains same in all inertial frames of reference.

If we measure only r^2 in frame S, and r'^2 in frame S' then the value of r^2 and r'^2 will be different in different inertial frames of references. On the other hand, if we measure $(ct)^2 - r^2$ its value is same in all inertial frames of references, i.e.,

$$r'^2 \neq r^2 \quad \text{not Lorentz invariant}$$

but

$$(ct)^2 - r^2 \text{ Lorentz invariant}$$

Now we know that

$$(ct)^2 - r^2 = 0$$
$$\text{or}$$

$$(ct)^2 - x^2 - y^2 - z^2 = 0 \tag{5.11}$$

The Lorentz invariant quantity mentioned in Eq. (5.11) has $+$ sign on first term but all other terms have a $-$ sign. But a dot product generates only $+$ signs. In 4D Euclidean space:

$$x_i \cdot x^i = x_1 x^1 + x_2 x^2 + x_3 x^3 + x_4 x^4 \tag{5.12}$$

By summation convention we mean that if a suffix occurs twice in a term, once in the upper position and once in the lower position, then that suffix implies sum over defined range. If the range is not given, then it is to be noted that the range is from 1 to 4.

5.3   Four-Dimensional (4D) Dot Product

In order to write Eq. (5.11) as 4D dot product we use matrix tensor $g_{\mu\nu}$

$$\begin{pmatrix} ct & x & y & z \end{pmatrix} \begin{pmatrix} 1 & 0 & 0 & 0 \\ 0 & -1 & 0 & 0 \\ 0 & 0 & -1 & 0 \\ 0 & 0 & 0 & -1 \end{pmatrix} = \begin{pmatrix} ct & -x & -y & -z \end{pmatrix} \tag{5.13}$$

By following Einstein's summation convention, the above equation in compact notation can be written as

$$x^\mu g_{\mu\nu} = x_\nu \tag{5.14}$$

where $X^\mu = x^\mu = \{ct, x, y, z\}$ is the contravariant vector with four components, whereas, $X_\mu = x_\mu = \{ct, -x, -y, -z\}$ is the covariant vector with four components. The quantity $g_{\mu\nu}$ is called the metric tensor. It is also called a fundamental tensor. It is a second rank covariant symmetric tensor. Thus, Eq. (5.11) can be written as follows:

$$X^\mu \cdot X_\mu = (ct)^2 - x^2 - y^2 - z^2 \tag{5.15}$$

Example 5.1 Consider two events in spacetime where the position and time coordinates of the events in the frame S are (t_1, x_1, y_1, z_1) and (t_2, x_2, y_2, z_2). Define the spacetime interval S^2 between these events by the expression.

$$S^2 = c^2(t_2 - t_1)^2 - (x_2 - x_1)^2 - (y_2 - y_1)^2 - (z_2 - z_1)^2$$

Prove that S^2 is Lorentz invariant.

Solution:

The space time interval in the given frame S is

$$S^2 = c^2(t_2 - t_1)^2 - (x_2 - x_1)^2 - (y_2 - y_1)^2 - (z_2 - z_1)^2$$

Let $\Delta t = (t_2 - t_1), \;\; \Delta x = (x_2 - x_1), \;\; \Delta y = (y_2 - y_1), \;\; \Delta z = (z_2 - z_1)$

$$S^2 = c^2(\Delta t)^2 - (\Delta x)^2 - (\Delta y)^2 - (\Delta z)^2$$

Consider the Lorentz transformation along x-axis where, v is the velocity between S and S'. The transformation equations are

$$t' = \gamma\left(t - \frac{vx}{c^2}\right)$$

$$x' = \gamma(x - vt)$$

$$y' = y$$

$$z' = z$$

Using the difference in coordinates, the time interval $\Delta t\prime$ and the spatial interval $\Delta x\prime$ in the frame S' become

$$\Delta t' = \gamma\left(\Delta t - \frac{v\Delta x}{c^2}\right)$$

$$\Delta x' = \gamma(\Delta x - v\Delta t)$$

$$\Delta y' = \Delta y$$

$$\Delta z' = \Delta z$$

$$S' = c^2\left(\Delta t'\right)^2 - \left(\Delta x'\right)^2 - \left(\Delta y'\right)^2 - \left(\Delta z'\right)^2$$

Here, $\gamma^2\left(c^2 - v^2\right) = c^2$

$$S\prime = c^2(\Delta t)^2 - (\Delta x)^2 - (\Delta y)^2 - (\Delta z)^2$$

Thus, S^2 is Lorentz invariant.

Example 5.2 Given two four vectors in spacetime $A^\mu = \left(A^0, A^1, A^2, A^3\right)$ and $B^\mu = \left(B^0, B^1, B^2, B^3\right)$ where $A^0 = ct_A$, $A^1 = x_A$, $A^2 = y_A$, $A^3 = z_A$ and similarly for B^μ. Define the four-dimensional dot product.

$$A_\mu B^\mu = A^0 B^0 - A^1 B^1 - A^2 B^2 - A^3 B^3$$

Show that this dot product remains invariant under Lorentz transformation.

Solution:

From the concept of four vectors, we write four-vector dot product as follows:

$$A_\mu B^\mu = A^0 B^0 - A^1 B^1 - A^2 B^2 - A^3 B^3$$

for a Lorentz transformation along the x-axis. Therefore, the transformation equations for A^0 and A^1 are

$$A^{0'} = \gamma\left(A^0 - \frac{vA^1}{c}\right)$$

$$A^{1'} = \gamma\left(A^1 - vA^0\right)$$

with $A^{2'} = A^2$ and $A^{3'} = A^3$
   Similarly, for B^μ

$$B^{0'} = \gamma\left(B^0 - \frac{vB^1}{c}\right)$$

$$B^{1'} = \gamma\left(B^1 - vB^0\right)$$

with $B^{2'} = B^2$ and $B^{3'} = B^3$
   Using the Lorentz transformation, the dot product of $A_{\mu'} B^{\mu'}$ in the S' frame is

$$A_{\mu'} B^{\mu'} = A^{0'} B^{0'} - A^{1'} B^{1'} - A^{2'} B^{2'} - A^{3'} B^{3'}$$

Substitute the transformed components

$$A_{\mu'} B^{\mu'} = \gamma^2\left(A^0 - \frac{vA^1}{c}\right)\left(B^0 - \frac{vB^1}{c}\right) - \gamma^2\left(A^1 - vA^0\right)\left(B^1 - vB^0\right)$$

Using $\gamma^2\left(1 - \frac{v^2}{c^2}\right) = 1$.
which substantiates that the product $A_\mu B^\mu$ is Lorentz invariant.

5.4 The Divergence

The divergence of any vector function is defined as follows:

$$\vec{\nabla} \cdot \vec{f} = \left(\frac{d}{dx}, \frac{d}{dy}, \frac{d}{dz} \right) \vec{f} \tag{5.16}$$

This represents the rate at which a vector field spreads out from a given point in 3D space.

Thus, to extend it into 4D Minkowski space.

In a compact notation, Eq. (5.16) can be written as

$$\vec{\nabla} \cdot \vec{f} = \frac{df^{\mu}}{dx^{\mu}} \tag{5.17}$$

If we follow the summation convention, the above equation is generalized as follows:

$$\vec{\nabla} \cdot \vec{f} = \partial_{\mu} f^{\mu} \tag{5.18}$$

where

$$\partial_{\mu} = \frac{d}{dx^{\mu}} = \left(\frac{1}{c} \frac{d}{d(ct)}, \frac{d}{d(x)}, \frac{d}{d(y)}, \frac{d}{d(z)} \right) \tag{5.19}$$

which is also called a covariant derivative

$$\partial_{\mu} = \left(\frac{1}{c} \frac{d}{dt}, \frac{d}{dx}, \frac{d}{dy}, \frac{d}{dz} \right)$$

and

$$\partial^{\mu} = \frac{d}{dx_{\mu}} = \left(\frac{d}{d(ct)}, \frac{d}{d(-x)}, \frac{d}{d(-y)}, \frac{d}{d(-z)} \right) \tag{5.20}$$

which is called the contravariant derivative.

Example 5.3 Consider a 4D vector field $F^{\mu} = \left(F^0, F^1, F^2, F^3 \right)$ in the Minkowski space, where $F^0 = \phi$ (a scalar field) and $F^i = \vec{A}$ (a 3D vector field with components A^x, A^y, A^z). Show that the divergence $\partial_{\mu} F^{\mu}$ is Lorentz invariant.

Solution:

The divergence of F^{μ} in 4D Minkowski space is given by

$$\partial_{\mu} F^{\mu} = \frac{\partial F^0}{\partial(ct)} + \frac{\partial F^1}{\partial x} + \frac{\partial F^2}{\partial y} + \frac{\partial F^3}{\partial z}$$

Let $F^0 = \phi$ a time-dependent scalar field and $F^1 = A_x$, $F^2 = A_y$ and $F^3 = A_z$ representing the spatial components of the vector $\vec{A}$.

$$\partial_\mu F^\mu = \frac{\partial \phi}{\partial (ct)} + \vec{\nabla} \cdot \vec{A}$$

Using Lorentz transformation in the x direction, we have

$$F^{0'} = \gamma \left(F^0 - \frac{vF^1}{c} \right)$$

$$F^{1'} = \gamma \left(F^1 - vF^0 \right)$$

with $F^{2'} = F^2$ and $F^{3'} = F^3$

The transformed derivatives are

$$\partial_{0'} = \gamma \left(\partial_0 + \frac{v}{c}\partial_1 \right), \quad \partial_{1'} = \gamma \left(\partial_1 + \frac{v}{c}\partial_0 \right)$$

$\partial_{2'} = \partial_2$ and $\partial_{3'} = \partial_3$

$$\partial_{\mu'} F^{\mu'} = \partial_{0'} F^{0'} + \partial_{1'} F^{1'} + \partial_{2'} F^{2'} + \partial_{3'} F^{3'}$$

Expanding each term using the transformation

$$\partial_{\mu'} F^{\mu'} = \gamma \left(\partial_0 F^0 - \frac{v\partial_1 F^1}{c} - \frac{v\partial_0 F^1}{c} + \frac{v^2 \partial_1 F^1}{c^2} \right) + \partial_1 F^1 + \partial_2 F^2 + \partial_3 F^3$$

Using $\gamma^2 \left(1 - \frac{v^2}{c^2} \right) = 1 = 1$

$$\partial_{\mu'} F^{\mu'} = \left(\partial_0 F^0 + \partial_1 F^1 + \partial_2 F^2 + \partial_3 F^3 \right)$$

5.5   Energy–Momentum Relation

The energy–momentum relation is given by

$$E^2 = p^2 c^2 + m^2 c^4 \tag{5.21}$$

The above relation is true for all particles in all inertial reference frames. The relativistic energy, E of a particle of rest mass m and momentum p is given by

$$\frac{E^2}{c^2} = p^2 + m^2 c^2 \tag{5.22}$$

$$\left(\frac{E}{c}\right)^2 - \left(p_x^2 + p_y^2 + p_z^2\right) = (mc)^2 \tag{5.23}$$

This result is tremendously useful. It helps us to calculate E provided we know p or calculate p provided we know E, without ever been knowing the velocity. From this expression we can recognize that:

$$X^{\mu} = \left(\frac{E}{C}, p_x, p_y, p_z\right) \quad \text{and} \quad X_{\mu} = \left(\frac{E}{C}, -p_x, -p_y, -p_z\right)$$

Thus Eq. (5.23) can be written as follows:

$$X_{\mu} X^{\mu} = (mc)^2 \tag{5.24}$$

5.6   The Continuity Equation

The conservation of charge demands that the charge density at any arbitrary point in space is related to the current density in that region by a continuity equation as follows:

$$\vec{\nabla} \cdot \vec{J} + \frac{d\rho}{dt} = 0 \tag{5.25}$$

This equation has immense physical significance. One of the important physical interpretations of this equation is that any decrease in charge inside a small volume with time must correspond to a flow of charge out through the surface encompassing that volume. This equation assumes the nice compact form when written in terms of J^{μ}.

$$\frac{1}{c}\frac{d(ct)}{dt} + \vec{\nabla} \cdot \vec{J} = 0 \tag{5.26}$$

$$\frac{1}{c}\frac{d(ct)}{dt} + \frac{dJ_x}{dx} + \frac{dJ_y}{dy} + \frac{dJ_z}{dz} = 0 \tag{5.27}$$

Then, the corresponding current density 4-vector are

$$X^{\mu} = \left(ct, J_x, J_y, J_z\right) = J^{\mu} \quad \text{and} \quad X_{\mu} = \left(ct, -J_x, -J_y, -J_z\right) = J_{\mu}$$

In compact form we write Eq. (5.27) as follows:

$$\partial_\mu J^\mu = 0 \tag{5.28}$$

The summation over μ is implied. This equation substantiates that the current density four-vector is divergenceless.

5.7   The Lorentz Gauge

The Coulomb Gauge is stated as $\vec{\nabla} \cdot \vec{A} = 0$.

It does not possess any time component; therefore, we can't define a 4-vector for it and hence we don't have any concept of Lorentz invariance for it.

The Lorentz Gauge is defined as

$$\frac{1}{c^2}\frac{\mathrm{d}\phi}{\mathrm{d}t} + \vec{\nabla} \cdot \vec{A} = 0 \tag{5.29}$$

$$\frac{1}{c}\frac{\mathrm{d}}{\mathrm{d}t}\left(\frac{\phi}{c}\right) + \frac{\mathrm{d}A_x}{\mathrm{d}x} + \frac{\mathrm{d}A_y}{\mathrm{d}y} + \frac{\mathrm{d}A_z}{\mathrm{d}z} = 0 \tag{5.30}$$

The physical dimensions of $\left[\frac{\phi}{c}\right] \sim \vec{A}$, i.e., of magnetic vector potential. So,

$$X^\mu = \left(\frac{\phi}{c}, A_x, A_y, A_z\right) = A^\mu$$

$$X^\mu = \left(\frac{\phi}{c}, -A_x, -A_y, -A_z\right) = A_\mu \quad \text{and} \quad \partial_\mu = \frac{\mathrm{d}}{\mathrm{d}x^\mu}$$

Therefore, Eq. (5.30) can be written in compact notation as follows:

$$\partial_\mu A^\mu = 0 \tag{5.31}$$

As long as Lorentz gauge is satisfied, the potentials will satisfy the inhomogeneous wave equation, with the source term on the right side.

5.8   The Lorentz Transformation

Lorentz transformation provides us an idea of relating the coordinates of an event in two inertial frames of reference. An event is a physical process that occurs at a specified location (x, y, z) and at a particular instant of time (t). Let (x, y, z, t) be the coordinates of an event E in an inertial frame of reference S and let the coordinates of the same event in other inertial frame of reference $\overline{S}$ be $(\overline{x}, \overline{y}, \overline{z}, \overline{t})$.

The Lorentz transformation equations of an event in two inertial reference frames S and $\overline{S}$ having relative motion between them are given as

$$\overline{x} = \gamma\left(x - \frac{v}{c}ct\right)$$
$$\overline{y} = y$$
$$\overline{Z} = z$$
$$\overline{ct} = \gamma\left(ct - \frac{v}{c}x\right)$$

(5.32)

For simplicity, we denote

$$x \to x^1, y \to x^2, z \to x^3, ct \to x^4$$

$$\overline{x} \to \overline{x}^1, \overline{y} \to \overline{x}^2, \overline{z} \to \overline{x}^3, \overline{ct} \to \overline{x}^0.$$

The set of Eqs. (5.32) can be re-written in matrix form as follows

$$\begin{pmatrix} \overline{ct} \\ \overline{x} \\ \overline{y} \\ \overline{z} \end{pmatrix} = \begin{pmatrix} \gamma & -\gamma\frac{v}{c} & 0 & 0 \\ -\gamma\frac{v}{c} & \gamma & 0 & 0 \\ 0 & 0 & 1 & 0 \\ 0 & 0 & 0 & 1 \end{pmatrix} \begin{pmatrix} ct \\ x \\ y \\ z \end{pmatrix}$$

$$\begin{pmatrix} \overline{x}^0 \\ \overline{x}^1 \\ \overline{x}^2 \\ \overline{x}^3 \end{pmatrix} = \begin{pmatrix} \gamma & -\gamma\frac{v}{c} & 0 & 0 \\ -\gamma\frac{v}{c} & \gamma & 0 & 0 \\ 0 & 0 & 1 & 0 \\ 0 & 0 & 0 & 1 \end{pmatrix} \begin{pmatrix} x^0 \\ x^1 \\ x^2 \\ x^3 \end{pmatrix}$$

(5.33)

In compact notation, we write

$$\overline{x}^\mu = \sum_{v=0}^{3} \lambda_v^\mu x^v$$

(5.34)

where λ_v^μ is the Lorentz transformation matrix. The superscript, μ labels the rows and the subscript, and v labels the columns of the matrix. One of the important property of writing equations in more abstract form is that these can be generalized to a situation where the relative motion is not along any common axis. The matrix λ would be more complicated but the structure of Eq. (5.34) remains unaltered. It is important to note here that the geometrical interpretation of Lorentz transformation is that it simply indicates the rotation in four-dimensional space.

For a particular case of transformation, where $\mu = 0$, the Eq. (5.34) would yield

$$\overline{x}^0 = \sum_{v=0}^{3} \lambda^0_v x^v$$

$$= \lambda^0_0 x^0 + \lambda^0_1 x^1 + \lambda^0_2 x^2 + \lambda^0_3 x^3$$

$$= \gamma(ct) + \left(-\gamma\frac{v}{c}\right)x + 0(y) + 0(x)$$

$$\overline{ct} = \gamma\left(ct - \frac{v}{c}x\right)$$

Example 5.4 Given two four vectors in spacetime $A^\mu = \left(A^0, A^1, A^2, A^3\right)$ and $B^\mu = \left(B^0, B^1, B^2, B^3\right)$ where $A^0 = ct_A$, $A^1 = x_A$, $A^2 = y_A$, $A^3 = z_A$ Define the four-dimensional dot product using the metric tensor $g_{\mu\nu}$.

$$A_\mu B^\mu = g_{\mu\nu} A^\mu B^\nu$$

Show that this dot product $A_\mu B^\mu$ remains invariant under Lorentz transformation.

Solution:

The metric tensor $g_{\mu\nu}$ in Minkowski space is defined by

$$g_{\mu\nu} = \begin{pmatrix} 1 & 0 & 0 & 0 \\ 0 & -1 & 0 & 0 \\ 0 & 0 & -1 & 0 \\ 0 & 0 & 0 & -1 \end{pmatrix}$$

Thus, the 4D dot product of A^μ and B^μ is

$$A_\mu B^\mu = g_{\mu\nu} A^\mu B^\nu = A^0 B^0 - A^1 B^1 - A^2 B^2 - A^3 B^3$$

$$A_\mu B^\mu = ct_A ct_B - x_A x_B - y_A y_B - z_A z_B$$

Using the Lorentz transformation along the x-axis with velocity v, the transformations for A^0 and A^1 are

$$A^{0'} = \gamma\left(A^0 - \frac{vA^1}{c}\right)$$

$$A^{1'} = \gamma\left(A^1 - vA^0\right)$$

where $\gamma = \frac{1}{\sqrt{1-\frac{v^2}{c^2}}}$ $A^{2'} = A^2$ and $A^{3'} = A^3$
Similarly, for B^μ

$$B^{0'} = \gamma\left(B^0 - \frac{vB^1}{c}\right)$$

$$B^{1'} = \gamma\left(B^1 - vB^0\right)$$

with $B^{2'} = B^2$ and $B^{3'} = B^3$

$$A_{\mu'}B^{\mu'} = g_{\mu\nu}A^{\mu'}B^{\nu'}$$

$$A_{\mu'}B^{\mu'} = A^{0'}B^{0'} - A^{1'}B^{1'} - A^{2'}B^{2'} - A^{3'}B^{3'}$$

Substitute the transformed components

$$A_{\mu'}B^{\mu'} = \gamma^2\left(A^0 - \frac{vA^1}{c}\right)\left(B^0 - \frac{vB^1}{c}\right) - \gamma^2\left(A^1 - vA^0\right)\left(B^1 - vB^0\right)$$

Using $\gamma^2\left(1 - \frac{v^2}{c^2}\right) = 1 = 1$

$$A_{\mu'}B^{\mu'} = A_{\mu}B^{\mu} = A^0B^0 - A^1B^1 - A^2B^2 - A^3B^3$$

5.9 Four-Vectors in Special Relativity

Tensors form the cornerstone of mathematical formulation of physical laws. Classical physics is concerned with the invariance of physical quantities under Galilean transformation and such transformations assume that the physical laws appear to be same to all observers which are stationary relative to each other. However, in the realm of theory of special relativity, it is widely known that the physical laws appear differently to observers that are in relative motion to each other, nonetheless, the physical laws must remain the same to all observers. The measurements of space and time intervals between two events made by one observer may differ from those of another observer, if they are in relative motion to each other. Transformation laws which relate the coordinates of one observer with those of another observer in uniform relative motion with respect to the first are referred as Lorentz transformations and these transformation laws include uniform relative motion in addition to translation and rotation. Thus, the notion of scalars and vectors must change accordingly. Further, for the mathematical description of physical laws, we cannot use those quantities that are not invariant. However, if we extend the set of transformations from Galilean to Lorentz transformations, we must incorporate all the inertial frames of reference. All the physical quantities that we come across in classical physics do not possess invariant magnitude and direction. The same concept, therefore, applies to scalar quantities. Thus, in order to be consistent with the theory of special relativity, we have to form the sets of four components that transform according to specified tensor laws under Lorentz transformation. In particular, the ordered sets of four elements transforming according to vector laws have been given special name of four-vectors or world vectors. It is pertinent to note here that these are not only four-dimensional vectors but they are the vectors of the Minkowski space, i.e., vectors possessing invariance under Lorentz transformation. A few such vectors are briefly discussed as follows:

5.9.1 *Four Vector*

We are conversant with the idea of four-dimensional space and hence it is logical to extend ordinary vector analysis (3 vectors) to four dimensions (i.e., 4 vectors). These four-dimensional vectors are four vectors or world vectors. A four-vector is a four-dimensional quantity that undergoes Lorentz transformation from one inertial frame to another inertial frame. It is worthwhile to mention here that the four-dimensional coordinates are orthogonal. We now define a contravariant 4-vector as any set of four components

$$a^\mu = \left(a^0, a^1, a^2, a^3\right)$$

A four vector maintains its form under Lorentz transformation. The transformation is given by

$$\bar{a}^\mu = \sum_{v=0}^{3} \lambda_v^\mu a^v \tag{5.35}$$

Four vectors are useful in expressing various physical quantities that are conserved in relativistic interactions, such as energy-momentum 4-vector and 4-current.

The length of the 4-vector remains invariant under Lorentz transformation which is demonstrated by dot product (inner product or Minkowski inner product).

$$a^\mu a_\mu = \left(a^0\right)^2 - \left(a^1\right)^2 - \left(a^2\right)^2 - \left(a^3\right)^2 \tag{5.36}$$

This property is crucial in relativistic physics, ensuring that quantities such as spacetime intervals and the energy-momentum 4-vector maintain their physical meaning in all inertial frames.

5.9.2 The Velocity Four-Vector

The velocity 4-vector is defined as

$$U^\mu = \frac{\mathrm{d}x^\mu}{\mathrm{d}\tau} = \left(\frac{\mathrm{d}x^0}{\mathrm{d}\tau}, \frac{\mathrm{d}x^1}{\mathrm{d}\tau}, \frac{\mathrm{d}x^2}{\mathrm{d}\tau}, \frac{\mathrm{d}x^3}{\mathrm{d}\tau}\right) \tag{5.37}$$

The infinitesimal displacement is given by

$$\mathrm{d}\bar{x}^\mu = \sum_{v=0}^{3} \lambda_v^\mu \mathrm{d}x^v \tag{5.38}$$

Equation (5.37) is obtained by differentiating x^μ with respect to the proper time but we want to calculate the velocity with respect to the time of an observer. Therefore, we require transformation equations that relate proper time with the observer's time. The time dilation expression is given by

$$\mathrm{d}t = \gamma \mathrm{d}\tau \tag{5.39}$$

Thus, Eq. (5.37) can be written as follows:

$$U^\mu = \left(\gamma \frac{c\mathrm{d}t}{\mathrm{d}t}, \gamma \frac{\mathrm{d}x}{\mathrm{d}t}, \gamma \frac{\mathrm{d}y}{\mathrm{d}t}, \gamma \frac{\mathrm{d}z}{\mathrm{d}t}\right)$$

$$U^\mu = \gamma\left(c, v_x, v_y, v_z\right)$$

It is evident from this expression that the three components of the four-velocity are the three components of the three-vector velocity times γ.

$$U^\mu = (\gamma c, \gamma \vec{v}) \tag{5.40}$$

Similarly, we can write

$$U_\mu = (\gamma c, -\gamma \vec{v}) \tag{5.41}$$

It is worthwhile to mention here that the norm—the magnitude or vector invariant length—of the four-velocity is not only unchanged but is same for all physical objects (matter plus energy).

Therefore, we have

$$U^\mu U_\mu = (\gamma c)^2 - (\gamma v)^2 = \gamma\left(c^2 - v^2\right) = \frac{c^2 - v^2}{c^2 - v^2} \times c^2$$

$$U^\mu U_\mu = c^2 = \text{constant} \tag{5.42}$$

which is the invariant quantity under Lorentz transformation.

5.9.3 *The Acceleration Four-Vector*

The acceleration 4-vector is established by differentiating velocity four-vector with respect to the proper time τ and is, therefore, given by

$$
\begin{aligned}
a^\mu &= \frac{\mathrm{d}u^\mu}{\mathrm{d}\tau} = \frac{\mathrm{d}}{\mathrm{d}\tau}\left(\frac{\mathrm{d}x^\mu}{\mathrm{d}\tau}\right) \\
&= \frac{\mathrm{d}}{\mathrm{d}\tau}\left(\gamma c,\, \gamma v_x,\, \gamma v_y,\, \gamma v_z\right) \\
&= \gamma\frac{\mathrm{d}}{\mathrm{d}\tau}\left(\gamma c,\, \gamma v_x,\, \gamma v_y,\, \gamma v_z\right) \\
&= \gamma\left(c\frac{\mathrm{d}\gamma}{\mathrm{d}t},\, \gamma\frac{\mathrm{d}v_x}{\mathrm{d}t},\, v_x\frac{\mathrm{d}\gamma}{\mathrm{d}t},\, \gamma\frac{\mathrm{d}v_y}{\mathrm{d}t},\, v_y\frac{\mathrm{d}\gamma}{\mathrm{d}t},\, \gamma\frac{\mathrm{d}v_z}{\mathrm{d}t},\, v_z\frac{\mathrm{d}\gamma}{\mathrm{d}t}\right)
\end{aligned}
\tag{5.43}
$$

$$
a^\mu = \gamma\left(c\frac{\mathrm{d}\gamma}{\mathrm{d}t},\, \gamma\frac{\mathrm{d}\vec{v}}{\mathrm{d}t},\, \vec{v}\frac{\mathrm{d}\gamma}{\mathrm{d}t}\right)
\tag{5.44}
$$

The acceleration 4-vector has a component that is parallel to the acceleration three-vector and a part which is parallel to the velocity three-vector.

5.9.4 **The Momentum Four-Vector**

The general extension of 3-vector momentum to 4-vector momentum follows from the transformation analysis and from the notion of how masses are transformed under such transformations. The 4-vector momentum is generally written as follows:

$$
\begin{aligned}
P^\mu &= m_0 u^\mu = m_0\gamma\left(c,\, v_x,\, v_y,\, v_z\right) \\
&= m_0\gamma\left(c,\, \vec{v}\right)
\end{aligned}
\tag{5.45}
$$

$$
P^0 = \frac{mc^2}{c} = \frac{E}{c}
\tag{5.46}
$$

$$
P^1 = m_0\gamma v_x = \frac{m_0}{\sqrt{1 - \frac{v^2}{c^2}}}v_x = mv_x = P_x
\tag{5.47}
$$

Similarly, we can prove that $P^2 = P_y$ and $P^3 = P^z P^3 = P^z$
Thus,

$$
F^\mu = \frac{\mathrm{d}p^\mu}{\mathrm{d}\tau} = \frac{\mathrm{d}p^\mu}{\mathrm{d}t}\frac{\mathrm{d}t}{\mathrm{d}\tau} = \gamma\frac{\mathrm{d}p^\mu}{\mathrm{d}t}
\tag{5.48}
$$

It is evident from the above expression that the three spatial components are just the Newtonian 3-momentum where the mass of the particle has been replaced by $m_0\gamma$.

5.9.5 The Force Four-Vector

The force-four vector is defined by the following expression

$$F^\mu = \frac{dp^\mu}{d\tau} = \frac{dp^\mu}{dt}\frac{dt}{d\tau} = \gamma\frac{dp^\mu}{dt} \tag{5.49}$$

However, momentum 4-vector is defined as $p^\mu = (mc, \vec{p})$
Therefore, Eq. (5.49) can be written as follows:

$$
\begin{aligned}
F^\mu &= \gamma\frac{d}{dt}(mc, \vec{p})\\
&= \gamma\left(c\frac{dm}{dt}, \frac{d\vec{p}}{dt}\right)\\
&= \left(\gamma c\frac{dm}{dt}, \gamma\vec{F}\right)\\
F^\mu &= \gamma\left(c\frac{dm}{dt}, \vec{F}\right)
\end{aligned}
\tag{5.50}
$$

It is worth noting that four-force could be time-like, space-like and null. If we get a frame of reference where the three-force acting on a particle is zero, however, the particle is interchanging internal energy with the surroundings, then the four-force is time-like. Otherwise, it is space-like.

5.10 The Lorentz Contraction and Time Dilation

According to theory of special relativity, the length of an object appears shorter only along the direction of its motion. The volume element in frame s is given by $dv = dxdydz$.

For an observer in frame s, the length appears to be contracted and the contraction in length is (Fig. 5.1)

$$dx' = \frac{dx}{\gamma} \tag{5.51}$$

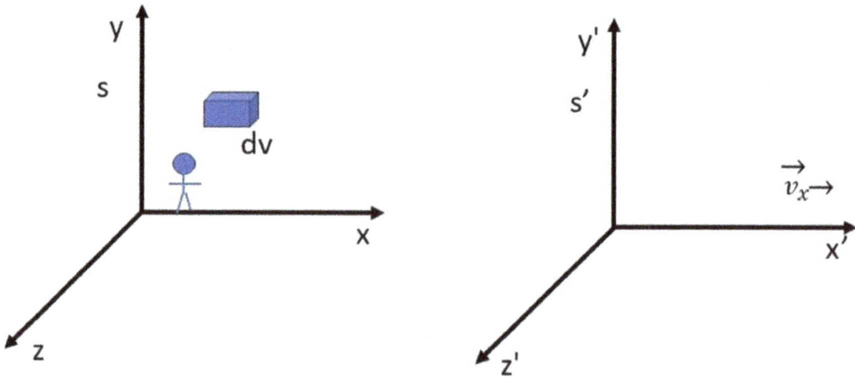

Fig. 5.1 Diagrammatic representation of two inertial frames of reference

For an observer in frame $s\prime$ the length of an object remains the same. Furthermore, the time interval between the events is dilated and the time dilation is

$$dt' = d\tau\gamma \tag{5.52}$$

Hence, we may say that a moving clock appears to go slow.
And relativistic mass is

$$M = \gamma m_0 \tag{5.53}$$

Since, volume element is not Lorentz invariant because the length changes along the direction of motion of an object. Therefore,

$$d\tau \neq d\tau' \tag{5.54}$$

5.11 The Transformation Equations for $\vec{E}$ and $\vec{B}$

It has been observed in various scenarios that the electric field appears to be a magnetic field for another observer. We are interested to know how the fields transform. Further, we assume that the transformation rules are unaltered irrespective of the fact how the fields are produced. Electric fields produced by changing magnetic fields transform in the same way as those set up by stationary charges.

(i) **Electric Field**: Let a frame $S\prime$ is moving with velocity v_x with respect to the frame s. Consider a parallel plate capacitor which is at rest in frame S and carries a charge density given by (Fig. 5.2)

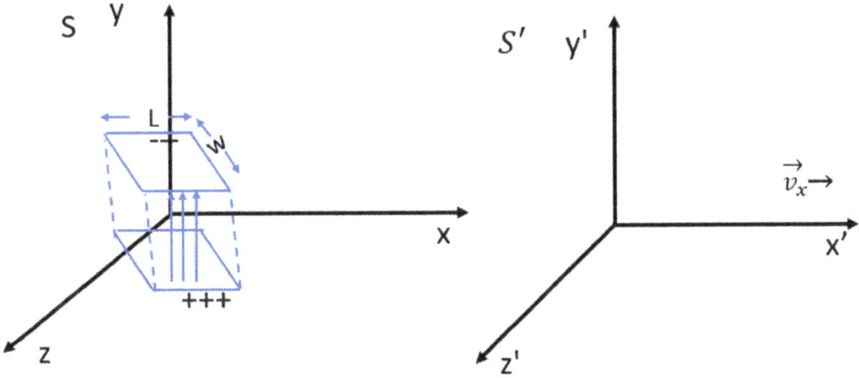

Fig. 5.2 Symbolic representation of an inertial frame S' moving with velocity $\vec{v}_x$ with respect to frame S

$$\sigma_{\text{plate}} = \frac{q}{A} \tag{5.55}$$

The area of the plates in inertial frame s is given by

$$\text{Area} = LW \tag{5.56}$$

The area of the plates in inertial frame S' which is moving with velocity v_x along x-axis is given by (Fig. 5.3)

$$\text{Area}' = A' = \frac{L}{\gamma}W \tag{5.57}$$

The charge density on the plates observed in inertial frame S' is given by

$$\sigma' = \frac{q}{A'} = \frac{\gamma q}{lw} = \gamma \sigma \tag{5.58}$$

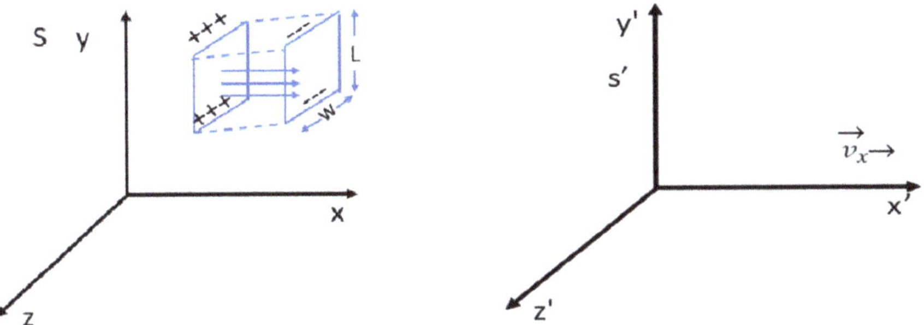

Fig. 5.3 Visualization of a parallel plate capacitor in an inertial frames of references

Accordingly,

$$\vec{E}' = \gamma \vec{E} \tag{5.59}$$

Therefore,

$$\vec{E}'(\perp_r) = \gamma \vec{E}_s(\perp_r) \tag{5.60}$$

The above equation pertains to the component of $\vec{E}$ that are produced perpendicular to the direction of motion of S. Since, no change occurs in parallel component of $\vec{E}$ as thickness decreases which has nothing to do with area of the plate. Therefore, it follows that

$$\vec{E}_{s'}^{\parallel} = \vec{E}_s^{\parallel} \tag{5.61}$$

Since, $\vec{v} = v_0 \hat{x}$. If the plates are lined up in xy-plane. Therefore, it will be the separation between the plates that is contracted whereas, L and W are the same. Hence, we may write

$$\vec{E} = E_1 \hat{x} + E_2 \hat{y} \tag{5.62}$$

$$\vec{E}' = E_1 \hat{x} + \gamma E_2 \hat{y} \tag{5.63}$$

(ii) **Magnetic Field**: Consider a solenoid aligned parallel to x-axis and is at rest in inertial frame s. The magnetic field within the coils is (Fig. 5.4)

$$\vec{B} = \mu_0 n I \vec{x} \tag{5.64}$$

where n is the number of turns per unit length and I is the current through the coils.

Due to the contraction of length in inertial frame $S\prime$, the number of turns increases. Therefore, it follows that

$$\vec{B}' = \mu_0 n' I \hat{x} \tag{5.65}$$

where

$$n\prime = \gamma n \tag{5.66}$$

Further, the time dilates. Therefore, the current (charge per unit time) in inertial frame s is given by

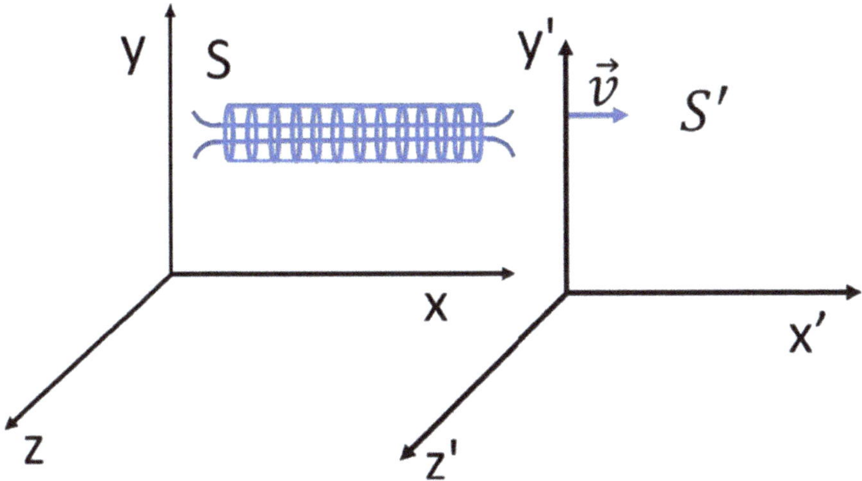

Fig. 5.4 Graphic representation of a solenoid in the inertial frames of references S and S'

$$I' = \frac{I}{\gamma} \tag{5.67}$$

Substitute Eqs. (5.66) and (5.67) in Eq. (5.65), we conclude that

$$\vec{B}' = \mu_0 \gamma n \frac{I}{\gamma} = \vec{B}' \tag{5.68}$$

Let us consider the following alignment of the solenoid. Thus, the magnetic field within the solenoid as observed in frame $s\prime$ is given by (Fig. 5.5)

$$\vec{B}' = \mu_0 I' n' \hat{y} \tag{5.69}$$

where as usual $I' = \frac{I}{\gamma}$

However, the circle has changed in to ellipse, therefore, the number of turns per unit length is

$$n' = \gamma^2 n \tag{5.70}$$

where r and ϕ changes as circle moves in $\vec{B}$. One γ comes from r and other from ϕ. Thus, Eq. (5.69) becomes

$$\vec{B}' = \mu_0 \gamma^2 n \frac{I}{\gamma} = \mu_0 n I \gamma = \gamma \vec{B} \tag{5.71}$$

Thus, from the above discussion it follows that

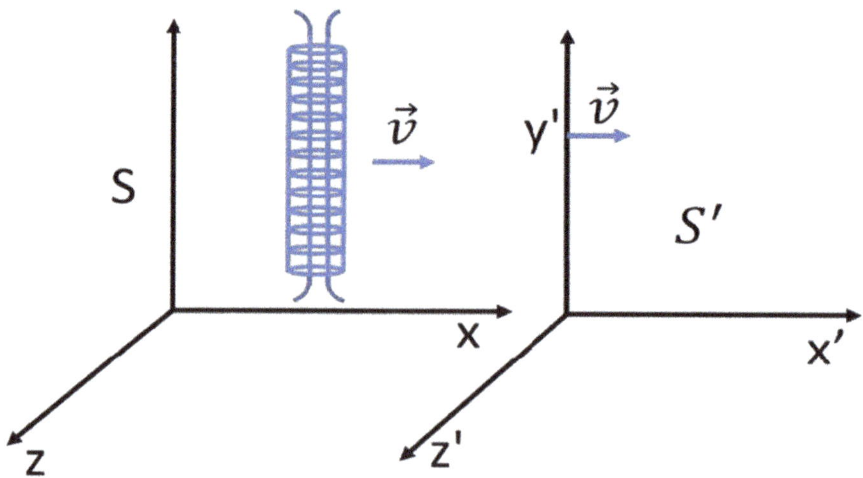

Fig. 5.5 Visualization of a solenoid placed parallel to Z-axis

$$\vec{B}_{S'}^{(\perp)} = \gamma \, \vec{B}_{S}^{(\perp)} \tag{5.72}$$

Hence, the entire set of transformation rules is enumerated as follows:

$$\vec{E}_{S'}^{(//)} = \vec{E}_{S}^{(//)}, \quad \vec{B}_{S'}^{(//)} = \vec{B}_{S}^{(//)} \tag{5.73}$$

$$\vec{E}_{S'}^{(\perp)} = \gamma E_{S}^{(\perp)}, \quad \vec{B}_{S'}^{(\perp)} = \gamma \vec{B}_{S}^{(\perp)}$$

5.12   Lorentz-Invariant Quantities

A quantity that does not undergo any change with the change in frame of reference is said to be Lorentz-invariant quantity. These quantities are, therefore, independent of the frame of reference. Further, it follows from this preposition that the laws of physics appear to be same for different observers irrespective of their state of motion. The various Lorentz invariant quantities include.

(1) Charge (q)
(2) Speed of light c
(3) $(ct)^2 - x^2 - y^2 - z^2$ (spacetime interval)
(4) $E^2 - c^2 B^2$
(5) $\vec{E} \cdot \vec{B}$
(6) $\nabla^2 - \dfrac{1}{C^2}\dfrac{d^2}{dt^2}$
(7) $E^2 - P^2 C^2$
(8) Maxwell's equation.

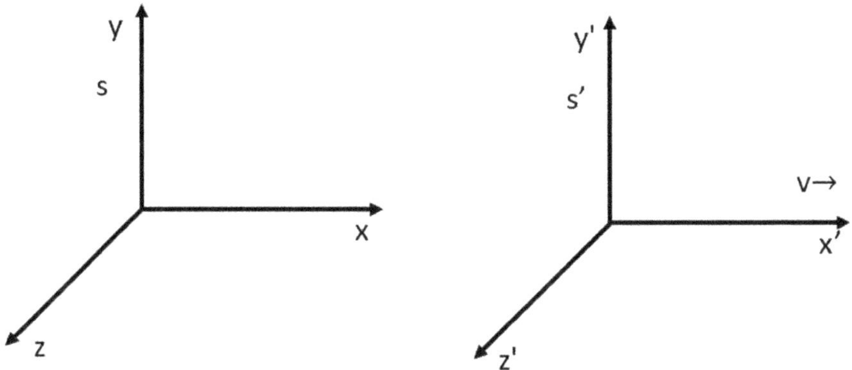

Fig. 5.6 Diagrammatic representation of the inertial frames of reference S and S'

Example 5.5 The electric field $E(r) = \frac{\lambda}{2\pi\varepsilon_0}\frac{x\hat{\imath}+y\hat{\jmath}}{x^2+y^2}$ due to charged infinite line along z-axis in inertial frame s. In a frame moving with a constant velocity w.r.t s along z axis, calculate $\vec{E}$ (Fig. 5.6)

Solution:

We need to find $\vec{E}'$

$$\vec{E}_z = \vec{E}_{z'}$$
$$\vec{E}_{x'} = \gamma\vec{E}_x \qquad\qquad (5.74)$$
$$\vec{E}_{y'} = \gamma\vec{E}_y$$

Example 5.6 In an inertial frame of reference, observer 'A' measures electric field as $E = (\alpha, 0, 0)$ and magnetic field $B = (\alpha, 0, 2\alpha)$ in a region where α is a constant. Another observer 'B' moving with constant velocity w.r.t A measures electric and magnetic fields respectively as: $\vec{E}' = \left(\vec{E}_{x'}, \alpha, 0\right)$; $\vec{B}' = \left(\alpha, \vec{B}_{y'}, \alpha\right)$. Determine what is the correct option for E'_x and B'_y from the following (Fig. 5.7).

(a) $-\frac{3a}{2}$ and $\frac{5a}{2}$.
(b) α and -2α.
(c) 2α and $-\alpha$.
(d) -2α and α.

Solution:

$\vec{E}\cdot\vec{B}$. will be same for both the inertial observers A and B, i.e.,

$$\vec{E}\cdot\vec{B} = \vec{E}\prime\cdot\vec{B}\prime \qquad\qquad (5.75)$$

and

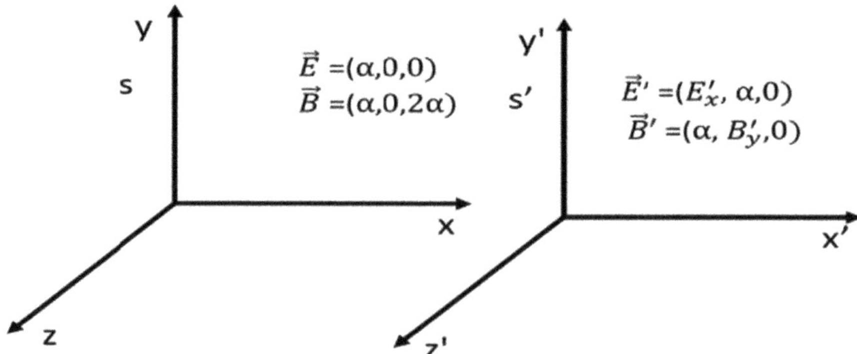

Fig. 5.7 Electric and magnetic field components in the inertial frames of references S and S'

$$E^2 - B^2 = E' - B'^2 \tag{5.76}$$

From Eq. (5.75), we can write

$$\vec{E} \cdot \vec{B} = \alpha^2$$
$$\vec{E}' \cdot \vec{B}' = \alpha E'_x + \alpha B'_y$$
$$E'_x + B'_y = \alpha \tag{5.77}$$

Further, from Eq. (5.76), we get

$$\alpha^2 - \alpha^2 - 4\alpha^2 = E'^2_x + \alpha^2 - \alpha^2 - B'^2_y$$

$$-4\alpha^2 = E'^2_x - B'^2_y \tag{5.78}$$

Equations (5.77) and (5.78) are only satisfied by option (a), i.e.,

$$E'_x = -\alpha \qquad\qquad B'_y = 2\alpha$$
$$A^2 = E'_x \alpha + \alpha B'_y \qquad -2\alpha^2 = \alpha^2 - 4\alpha^2 = -2\alpha^2$$
$$= -\alpha^2 + 2\alpha^2 = \alpha^2$$

Example 5.7 In an inertial frame uniform $\vec{E}$ and $\vec{B}$ are perpendicular to each other and satisfy $\left|\vec{E}\right|^2 - \left|\vec{B}\right|^2 = 29$. In another inertial frame which is moving with a constant velocity w.r.t. the 1st frame, the magnetic field is $2\sqrt{5}\hat{k}$ in second frame and E consistent with the previous observer is:

Solution:

From the transformation laws, we can write

$$\left|\vec{E}\right|^2 - \left|\vec{B}\right|^2 = \left|\vec{E'}\right|^2 - \left|\vec{B'}\right|^2$$

$$29 = \left|\vec{E'}\right|^2 - 20$$

$$\left|\vec{E'}\right|^2 = 29 + 20 = 49$$

$$\left|\vec{E'}\right| = \frac{7}{\sqrt{2}} \, (\hat{\imath} + \hat{\jmath}) \quad \text{It will be } \perp \text{ to } \vec{B}$$

5.13 Self-scalar Product

(i) **Velocity**: For $3 + 1$ dimensions the magnitude or norm is obtained from the self or dot product which has the same signature as the metric. The self-product of velocity four-vector is therefore, written as

$$V^\mu = \left(\gamma^C, \gamma^V\right); \; V_\mu = \left(\gamma^C, -\gamma^V\right)$$
$$V^\mu V_\mu = \left(\gamma^C, \gamma^V\right)\left(\gamma^C, -\gamma^V\right) = (\gamma c)^2 - (\gamma v)^2$$
$$= \gamma^2\left(c^2 - v^2\right)$$
$$= \gamma^2 c^2\left(1 - \frac{v^2}{c^2}\right)$$
$$= \left(\frac{\gamma^2 c^2}{\gamma^2}\right) = c^2 \tag{5.79}$$

Self-scalar product of velocity-four vector is c^2 (scalar). Further, we can conclude that every object moves with a $4 -$ velocity of magnitude c and the only effect of the Lorentz transformations is to change the direction of motion.

(ii) **Four-Momentum with Four-Velocity**

The concepts of four-momentum and four-force are fundamental in describing the conservation laws and dynamics of particles in special relativity. They provide a consistent way to express energy and momentum in four-dimensional spacetime.

The four-momentum in terms of four-velocity are given by the following expression

$$P^\mu = m_0 V^\mu \tag{5.80}$$

$$P^\mu = (mc, p) \tag{5.81}$$

The four-velocity V^μ of a particle is a four-vector that represents the rate of change of the particle's position in spacetime with respect to its proper time, τ. Proper time is the time measured in the frame of reference moving with the particle, making it an intrinsic property of the particle's motion. Defined in terms of the particle's spacetime coordinates $x^\mu = \{ct, x, y, z\}$ the four-velocity is given by:

$$V_\mu = (\gamma c, -\gamma v) \tag{5.82}$$

From Eqs. (5.80) and (5.82), we can write

$$P^\mu V_\mu = m_0 V^\mu V_\mu$$

Or, we can write

$$P^\mu V_\mu = m_0 c^2$$

This expression represents the scalar product of four-momentum and four-velocity. The term γmc^2 corresponds to the total relativistic energy of the particle, while γmv^2 captures the kinetic contribution to the energy in the spatial direction.

Further, from Eqs. (5.81) and (5.82), we can write

$$P^\mu V_\mu = (mc, p)(\gamma c, -\gamma v) = \gamma mc^2 - \gamma mv^2$$
$$= \gamma mc^2 \left(1 - \frac{V^2}{c^2} \right)$$
$$= \left(\frac{c^2 \gamma m}{\gamma^2} \right) = m_0 c^2 \tag{5.83}$$

This demonstrates that the invariant rest mass energy $m_0 c^2$ is a fundamental quantity, conserved across all inertial frames.

(iii)   Four-Force and Four-Velocity

In the framework of special relativity, four-velocity and four-force are critical concepts that extend the classical notions of velocity and force to four-dimensional spacetime. These four-dimensional vectors (or four-vectors) allow us to analyze the dynamics of objects moving at relativistic speeds, where the effects of special relativity are significant.

Four-force F^μ is the four-dimensional generalization of the classical force and describes how four-momentum changes with respect to proper time. It is defined as:

$$F^\mu = \left(\gamma c \frac{dm}{dt}, \gamma F \right)$$

And the four-velocity is given by:

$$V_\mu = (\gamma c, -\gamma v)$$

$$F^\mu V_\mu = \frac{\mathrm{d}P^\mu}{\mathrm{d}t}$$

where $P^\mu = m_0 V^\mu$

$$F^\mu V_\mu = \frac{\mathrm{d}V^\mu m_0}{\mathrm{d}t}$$

$$V_\mu F^\mu = \frac{\mathrm{d}V^2 m_0}{\mathrm{d}t^2}$$

Since the four-velocity and four-force are orthogonal, we have:

$$F^\mu V_\mu = 0 \tag{5.84}$$

This property reflects that in the particle's instantaneous rest frame, where $V^\mu = (c, 0, 0, 0)$, the force acts only in spatial directions without affecting the temporal component of four-velocity.

Further, we can write

$$F^\mu V_\mu = \left(\gamma c \frac{\mathrm{d}m}{\mathrm{d}t}, \gamma F \right)(\gamma c, -\gamma v)$$

$$0 = \left(\gamma^2 c^2 \frac{\mathrm{d}m}{\mathrm{d}t} - \gamma^2 F v \right)$$

$$c^2 \frac{\mathrm{d}m}{\mathrm{d}t} - F v = 0$$

$$c^2 \frac{\mathrm{d}m}{\mathrm{d}t} = F v$$

$$c^2 \mathrm{d}m = F v \mathrm{d}t$$

At $t = 0$, $m = m_0$. Integrating above equation, we get

$$\int_{m_0}^{m} c^2 \mathrm{d}m = \int_0^t F v \mathrm{d}t$$

$$\int_0^t F v \mathrm{d}t = c^2 (m - m_0) \tag{5.85}$$

Let us evaluate the integral on the L.H.S, we get

$$\int_0^t m\frac{dv}{dt} \cdot vdt = \int_0^t m\left(\frac{dv^2}{2dt}\right)dt$$

$$\int_0^t \frac{dmv^2}{dt2}dt = \frac{1}{2}mv^2$$

(5.86)

This equation indicates that the classical kinetic energy $\frac{1}{2}m \cdot v^2$ is the difference between the total relativistic energy and the rest energy. It represents the extra energy the particle gains due to its motion.

From Eqs. (5.85) and (5.86), we get

$$\frac{1}{2}mv^2 = c^2(m - m_0)$$

$$K \cdot E = c^2(m - m_0)$$

This expression shows that the kinetic energy depends on the difference between the relativistic mass m and the rest mass m_0 scaled by the square of the speed of light c^2.

$$Mc^2 = K \cdot E + m_0c^2$$

This relationship unites the rest energy m_0c^2 with the kinetic energy to yield the total energy of the particle in its moving frame.

$$E = T + m_0c^2$$

(5.87)

where T represents the relativistic kinetic energy of the particle. This equation reinforces that E, the total energy, combines both the particle's inherent rest energy and the additional energy it gains through motion.

5.14 Combined Electric and Magnetic Field Transformations

The transformation of electric and magnetic fields can be obtained while utilizing the Lorentz force law as the definition of $\vec{E}$ and $\vec{B}$. Let us consider two inertial frames S and S' as shown in below (Fig. 5.8), where S' moves with a constant velocity relative to S along the x-axis. The electric and magnetic fields in the original frames S are given by:

$$\vec{E} = E_x\hat{x} + E_y\hat{y} + E_z\hat{z}$$

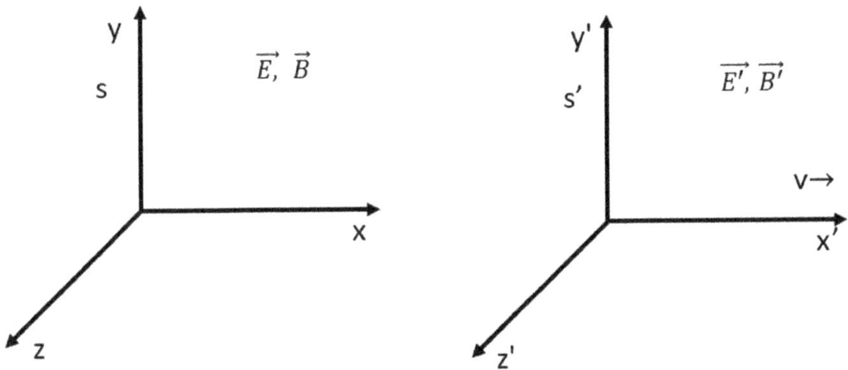

Fig. 5.8 Graphical representation of transformed vectors $\vec{E}$ and $\vec{B}$ in two inertial frames of reference S and S'

$$\vec{B} = B_x\hat{x} + B_y\hat{y} + B_z\hat{z}$$

(i) **Electric Field**: Let the electron is at rest with respect to the frame s and we assume it moves with velocity v along x'-axis relative to frame. s'. From the transformation equations of force, it follows that

$$\vec{E}_x = \vec{E}_{x'}$$
$$\vec{E}_{y'} = \gamma\left(\vec{E}_y - v\vec{B}_z\right)$$
$$\vec{E}'_y = \gamma\left(\vec{E}_y\right)\quad\left(\text{if } v\vec{B}z = 0\right)\quad\text{(Anticlockwise positive sign)}$$
$$\vec{E}'_z = \gamma\left(\vec{E}_z + v\vec{B}y\right)$$

(ii) **Magnetic Field**: It is pertinent to mention here that the electron is at rest in frame s, therefore, the magnetic field does not produce a force. We can, therefore, write the transformation law for the magnetic field as follows

$$\vec{B}'_x = \vec{B}_x$$

$$\vec{B}'_y = \gamma\left(\vec{B}_y - \frac{v}{C^2}\vec{E}z\right)$$

$$\vec{B}'_Z = \gamma\left(\vec{B}z - \frac{v}{C^2}\vec{E}y\right)$$

Motion along y direction

$$\vec{B}'_y = \vec{B}_y$$

$$\vec{B}z = \gamma\left(\vec{B}_Z - \frac{v}{C^2}\vec{E}x\right)$$

$$\vec{B}'_x = \gamma\left(\vec{B}_x - \frac{v}{C^2}\vec{E}z\right)$$

These equations describe how the electric and magnetic fields transform between two inertial frames moving relative to each other with velocity v.

Example 5.8 Using the Lorentz transformation matrix, show that for a particle moving along the x-axis, the transformation preserves the 4-vector form. Use the matrix form to find the transformed time and spatial coordinates (ct', x') of an event in S' given $ct = 4sc$, $x = 3m$, $v = 0.6c$ and verify using the Lorentz transformation.

Solution:

For motion along x-axis the Lorentz transformation matrix is

$$\lambda = \begin{pmatrix} \gamma & -\gamma\frac{v}{c} & 0 & 0 \\ -\gamma\frac{v}{c} & \gamma & 0 & 0 \\ 0 & 0 & 1 & 0 \\ 0 & 0 & 0 & 1 \end{pmatrix}$$

where $\gamma = \dfrac{1}{\sqrt{1-\frac{v^2}{c^2}}} = \dfrac{1}{\sqrt{1-(0.6)^2}} = 1.25$

$$X = \begin{pmatrix} ct \\ x \\ y \\ z \end{pmatrix} = \begin{pmatrix} 4cs \\ 3m \\ 0 \\ 0 \end{pmatrix}$$

Use the transformation $X' = \lambda X = \begin{pmatrix} \gamma & -\gamma\frac{v}{c} & 0 & 0 \\ -\gamma\frac{v}{c} & \gamma & 0 & 0 \\ 0 & 0 & 1 & 0 \\ 0 & 0 & 0 & 1 \end{pmatrix}\begin{pmatrix} 4cs \\ 3m \\ 0 \\ 0 \end{pmatrix}$

$$ct' = \gamma(4cs) - \gamma\frac{v}{c}(3m) = 2.75cs$$

$$x' = -\gamma\frac{v}{c}(4c) + \gamma(3m) = 0$$

The transformed coordinates are $ct' = 2.75cs$ and $x' = 0$, which verify that the Lorentz transformation preserves the 4-vector form.

Example 5.9 An electric field $\vec{E} = 500N/C$ is perpendicular to the direction of motion in a parallel plate capacitor, if an observer in frame S sees this electric field, calculate the electric field E' observed in a frame $S\prime$ moving with speed $v = 0.6c$ relative to S along the x-axis.

Solution:

Since the electric field is perpendicular to the direction of motion along the y-axis, the component transformation rule for the perpendicular electric field is

$$E'_\perp = \gamma E$$

$$\gamma = \frac{1}{\sqrt{1 - \frac{v^2}{c^2}}} = \frac{1}{\sqrt{1 - (0.64)^2}} = \frac{5}{4}$$

$$E' = \gamma E = \frac{5}{4}(500N/C) = 625N/C$$

The electric field observed in the frame S' is

$$E' = 625N/C$$

Example 5.10 In an inertial frame S, an observer measures an electric field $E = 6N/C$ and a magnetic field $B = 2 \times 10^{-8}$ T in a specific region of space. The frame S' moves with a velocity $v = 0.6\,c$ along the x-axis relative to S. Calculate the value of the Lorentz invariant quantity $E^2 - c^2 B^2$ in both frames and verify that it remains unchanged.

Solution:

In frame S

$$E^2 - c^2 B^2 = (6N/C)^2 - \left(3 \times 10^8 \text{m/s}\right)^2 \times \left(2 \times 10^{-8} \text{ T}\right)^2$$

Simplifying

$$E^2 - c^2 B^2 = 0$$

The Lorentz-invariant quantity $E^2 - c^2 B^2$ remains zero in both frames demonstrating its invariance under Lorentz transformation.

Unsolved Problems:

Problem 5.1 Given the electromagnetic tensor $F^{\mu\nu}$ in frame S.

$$F^{\mu\nu} = \begin{pmatrix} 0 & -E_x/c & -E_y/c & -E_z/c \\ E_x/c & 0 & -B_z & B_y \\ E_y/c & B_z & 0 & -B_x \\ E_z/c & -B_y & B_x & 0 \end{pmatrix}$$

Show that the quantity $F^{\mu\nu}F_{\mu\nu}$ is Lorentz invariant.

Problem 5.2 Consider two particles with 4-momenta $P_1^{\mu} = \left(\frac{E_1}{c}, p_{x1}, p_{y1}, p_{z1}\right)$ and $P_{2\mu} = \left(\frac{E_2}{c}, p_{x2}, p_{y2}, p_{z2}\right)$. Show that the quantity $P_1^{\mu} P_{2\mu}$ is invariant under Lorentz transformation.

Problem 5.3 Suppose $J^{\mu} = \left(\rho c, J_x, J_y, J_z\right)$ represents a 4-current density vector, where ρ is the charge density and J is the spatial current density. Show that the divergence of J^{μ}, defined as $\partial_{\mu}J^{\mu} = 0$ leads to the continuity equation $\frac{\partial \rho}{\partial t} + \vec{\nabla} \cdot \vec{J} = 0$.

Problem 5.4 For a particle of mass m and momentum p, the relativistic energy is given by $E = \sqrt{(pc)^2 + \left(mc^2\right)^2}$. Verify that this expression satisfies the relation $E^2 - (pc)^2 = \left(mc^2\right)^2$.

Problem 5.5 Consider a source emitting light with 4-momentum $P^{\mu} = \left(\frac{E}{c}, p_x, p_y, p_z\right)$ in frame S. If an observer in frame S' moves with velocity v along the x-axis relative to S. Show that the observed frequency v' of the light in S' is given by

$$v\prime = v\sqrt{\frac{1-\beta}{1+\beta}}$$

Problem 5.6 Let $U^{\mu} = \gamma\left(c, v_x, v_y, v_z\right)$ represents the velocity 4-vector of a particle moving with speed v in the x-direction. Show that the time component of U^{μ} corresponds to the time dilation factor.

Problem 5.7 Consider an electric field E along the y-axis and a magnetic field B along the z-axis in a frame S. Find the transformed electric field E'_y and the magnetic field B'_z in frame S', which moves along the x-axis with velocity v.

Ans. $E'_y = \frac{E - vB}{\sqrt{1 - \frac{v^2}{c^2}}}$, $B'_z = \frac{B - \frac{v}{c^2}E}{\sqrt{1 - \frac{v^2}{c^2}}}$.

Problem 5.8 Show that the 4-acceleration $A^{\mu} = \frac{dU^{\mu}}{d\tau}$ of a particle is orthogonal to its 4-velocity U^{μ}, where U^{μ} is the 4-velocity and τ is the proper time.

5.15 Summary

- **Introduction to Tensors**: Tensors generalize vectors and are essential for describing physical phenomena in non-relativistic and relativistic physics. They provide a mathematical framework for understanding physical laws under various transformations.
- **Lorentz-Invariant Quantities**: Quantities like $(ct)^2 - r^2$, energy–momentum relations and Maxwell's equations remain unchanged under Lorentz transformations, ensuring consistency across inertial frames.

- **Four-Dimensional Dot Product**: The scalar product in 4D spacetime uses the metric tensor $g_{\mu\nu}$, extending classical vector operations to relativistic contexts.
- **Energy–Momentum Relation:** Describes the link between energy, momentum and mass through $E^2 = p^2c^2 + m^2c^4$ fundamental for relativistic particle dynamics.
- **Continuity Equation**: Charge conservation is expressed as $\partial_\mu J^\mu = 0$, representing the divergence-free nature of current density in 4D spacetime.
- **Lorentz Gauge**: Compactly represented as $\partial_\mu A^\mu$, the Lorentz gauge ensures potentials satisfy the inhomogeneous wave equation with source terms.
- **Lorentz Transformation**: Relates spacetime coordinates between inertial frames via transformation matrices, interpreted geometrically as 4D rotations.
- **Four-Velocity and Four-Acceleration**: Four-velocity $U^\mu = \gamma(c, \vec{v})$ maintains an invariant magnitude c^2 under Lorentz transformations. Four-acceleration incorporates relativistic corrections, bridging classical and relativistic mechanics.
- **Four-Momentum and Four-Force**: Four-momentum $P^\mu = \gamma U^\mu$ encapsulates energy and momentum in 4D spacetime. Four-force $F^\mu = \frac{\mathrm{d}P^\mu}{\mathrm{d}\tau}$ governs relativistic dynamics.
- **Electric and Magnetic Field Transformations**: Electric and magnetic fields transform consistently across frames, e.g., $\vec{E}'_\perp = \gamma \vec{E}_\perp$, maintaining physical laws.
- **Lorentz Contraction and Time Dilation**: Relativistic effects include contracted lengths and dilated time intervals, critical for high-speed phenomena.
- **Self-scalar Products:** Invariant magnitudes for velocity, momentum and force four-vectors demonstrate the consistency of relativistic formulations.
- **Applications of Lorentz Invariance**: Spacetime intervals $(S)^2$, dot products and fields $\left(E^2 - c^2B^2\right)$ remain invariant under Lorentz transformations. Transformation rules are applied to scenarios like moving capacitors and solenoids.
- **Electric and Magnetic Field Relationships**: Field transformations illustrate how electric fields transform into magnetic fields under motion, adapting to relativistic velocities.
- **Relativistic Kinetic Energy**: The relationship $KE = c^2(m - m_0)$ integrates relativistic mass changes and rest energy, uniting classical and relativistic energy concepts.
- **Four-Force and Four-Velocity**: The orthogonality of four-force and four-velocity ensures energy conservation and captures interactions in spacetime dynamics.
- **Transformation of Combined Fields**: Electric and magnetic fields transform consistently, preserving Lorentz invariance and adhering to Maxwell's equations.

This chapter bridges classical mechanics and special relativity, providing mathematical tools and physical insights for analyzing relativistic electrodynamic systems.

Appendix

1. Fundamental Constants

Constant	Symbol	Value	Units	Description
Speed of light	c	2.998×10^8	m/s	Speed of light in vacuum
Elementary charge	e	1.602×10^{-19}	C	Charge of single electron
Permittivity of free space	ε_0	8.85×10^{-12}	F/m	Electric force in vacuum
Permeability of free space	μ_0	$4\pi \times 10^{-7}$	N/A^2	Magnetic force in vacuum
Electron mass	m_e	9.109×10^{-31}	kg	Mass of electron
Coulomb's constant	k_e	8.987×10^9	N-m^2/C^2	Electric force constant in vacuum

2. Mathematical Formulae

Formula	Expression	Description
Gradient	$\vec{\nabla} f = \left(\frac{\partial f}{\partial x}\right)\hat{x} + \left(\frac{\partial f}{\partial y}\right)\hat{y} + \left(\frac{\partial f}{\partial z}\right)\hat{z}$	Measures the rate and direction of change in a scalar field
Divergence	$\vec{\nabla}.\vec{A} = \left(\frac{\partial A_x}{\partial x}\right) + \left(\frac{\partial A_y}{\partial y}\right) + \left(\frac{\partial A_z}{\partial z}\right)$	Measures the net flow out of a point in a vector field
Curl	$\vec{\nabla} \times \vec{A} = \hat{x}\left(\frac{\partial V_z}{\partial y} - \frac{\partial V_y}{\partial z}\right)$ $-\hat{y}\left(\frac{\partial V_z}{\partial x} - \frac{\partial V_x}{\partial z}\right)$ $+\hat{z}\left(\frac{\partial V_y}{\partial x} - \frac{\partial V_x}{\partial y}\right)$	Describes rotation in a vector field, generating the field's circulation
Laplacian	$\nabla^2 f = \frac{\partial^2 f}{\partial x^2} + \frac{\partial^2 f}{\partial y^2} + \frac{\partial^2 f}{\partial z^2}$	Describes second derivative measures for fields

(continued)

© The Editor(s) (if applicable) and The Author(s), under exclusive license
to Springer Nature Singapore Pte Ltd. 2025
R. Ahmad Parra et al., *A Systematic Approach to Electrodynamics*,
https://doi.org/10.1007/978-981-96-3408-8

(continued)

Formula	Expression	Description
Divergence theorem	$\int_V \vec{\nabla} \cdot \vec{A}\, dV = \oint_S \vec{A} \cdot d\vec{s}$	Relates the flow across a surface to the divergence within a volume
Stokes' theorem	$\int_S \left(\vec{\nabla} \times \vec{A} \right) \cdot d\vec{s} = \oint_C \vec{A} \cdot d\vec{l}$	Relates the circulation around a loop to the curl over a surface
Fourier transform	$\mathcal{F}[f(x)] = \int_{-\infty}^{\infty} f(x) e^{-i2\pi kx} dx$	Converts functions from spatial to frequency domain
Gaussian integral	$\int_{-\infty}^{\infty} e^{-\alpha x^2}\, dx = \sqrt{\frac{\pi}{\alpha}}$	Evaluates integrals of the exponential functions common in physics

3. Basic Equations in Electrodynamics

Equation name	Expression	Description
Coulomb's law	$\vec{E} = \frac{1}{4\pi\varepsilon_0} \frac{\hat{r}}{r^2}$	Describes electric field from a point charge
Gauss's law (electrostatics)	$\vec{\nabla} \cdot \vec{E} = \frac{\rho}{\varepsilon_0}$	Relates electric field divergence to charge density
Faraday's law	$\vec{\nabla} \times \vec{E} = -\frac{\partial \vec{B}}{\partial t}$	Describes how a time-varying magnetic field induces an electric field
Ampere's law	$\vec{\nabla} \times \vec{B} = \mu_0 \vec{J} + \mu_0 \varepsilon_0 \frac{\partial \vec{E}}{\partial t}$	Describes how current and a time-varying electric field induce a magnetic field
Gauss's law (magnetostatics)	$\vec{\nabla} \cdot \vec{B} = 0$	Indicates that no magnetic monopoles exist
Lorentz force law	$\vec{F} = q \left(\vec{E} + \vec{v} \times \vec{B} \right)$	Describes force on a moving charge in electric and magnetic field
Poisson's equation	$\nabla^2 \phi = -\frac{\rho}{\varepsilon_0}$	Relates electric field to charge density
Wave equation	$\nabla^2 \vec{E} - \frac{1}{c^2} \frac{\partial^2 \vec{E}}{\partial t^2} = 0$	Describes propagation of electromagnetic waves
Continuity equation	$\vec{\nabla} \cdot \vec{J} + \frac{\partial \rho}{\partial t} = 0$	Represents conservation of charge
Maxwell's equations	$\vec{\nabla} \cdot \vec{E} = \frac{\rho}{\varepsilon_0}$ $\vec{\nabla} \cdot \vec{B} = 0$ $\vec{\nabla} \times \vec{E} = -\frac{\partial \vec{B}}{\partial t}$ $\vec{\nabla} \times \vec{B} = \mu_0 \vec{J} + \mu_0 \varepsilon_0 \frac{\partial \vec{E}}{\partial t}$	Set of four fundamental equations in classical electrodynamics

4. **Double Factorial**: The double factorial of an integer n, denoted by $n!!$ is the product of all the integers from n down to 1 that have the same parity (even or odd) as n.

(i) **Double Factorial (odd n)**

$$(2n - 1)!! = (2n - 1) \cdot (2n - 3) \ldots 3.1$$

(ii) **Double Factorial (even n)**

$$(2n)!! = (2n) \cdot (2n - 3) \ldots 4.2$$

(iii) **Recursive Formula**

$$(n)!! = n \cdot (n - 2)!!$$

(iv) $n! = (n)!! \cdot (n - 1)!!$ **for positive integer n**

(v) **Gamma Functions Relations (for even/odd cases)**

For odd integers $n = 2m + 1$

$$(2m + 1)!! = \frac{(2m + 1)!}{2^m \cdot m!}$$

For even integers $n = 2m$

$$(2m)!! = \frac{(2m)!}{2^m \cdot m!}$$

5. **Fourier Series**: The Fourier series decomposes a periodic function $f(x)$ into a sum of sines and cosines representing the function as an infinite series.

The general form with period 2π

$$f(x) = a_0 + \sum_{n=1}^{\infty} (a_n \cos(nx) + b_n \sin(nx))$$

where $a_0 = \frac{1}{2\pi} \int_{-\pi}^{\pi} f(x)\mathrm{d}x, \quad a_n = \frac{1}{\pi} \int_{-\pi}^{\pi} f(x) \cos(nx)\mathrm{d}x$

$$b_n = \frac{1}{\pi} \int_{-\pi}^{\pi} f(x) \sin(nx) dx$$

6. Fourier Series for Even and Odd Functions

(i) If $f(x)$ is even, all $b_n = 0$ and only cosine terms remain

$$f(x) = a_0 + \sum_{n=1}^{\infty} a_n \cos(nx)$$

(ii) If $f(x)$ is odd, all $a_n = 0$ and only sine terms remain

$$f(x) = \sum_{n=1}^{\infty} b_n \sin(nx)$$

7. Vector Calculus Operators in Cylindrical and Spherical Coordinates

Operator	Cylindrical Coordinates (r, ϕ, z)	Spherical Coordinates (r, θ, ϕ)
Gradient $\vec{\nabla} f$	$\frac{\partial f}{\partial r}\hat{r} + \frac{1}{r}\frac{\partial f}{\partial \phi}\hat{\phi} + \frac{\partial f}{\partial z}\hat{z}$	$\frac{\partial f}{\partial r}\hat{r} + \frac{1}{r}\frac{\partial f}{\partial \theta}\hat{\theta} + \frac{1}{r\sin(\theta)}\frac{\partial f}{\partial \phi}\hat{\phi}$
Divergence $\vec{\nabla} \cdot \vec{A}$	$\frac{1}{r}\frac{\partial(rA_r)}{\partial r} + \frac{1}{r}\frac{\partial(A_\phi)}{\partial \phi} + \frac{\partial A_z}{\partial z}$	$\frac{1}{r^2}\frac{\partial(r^2 A_r)}{\partial r} + \frac{1}{r\sin(\theta)}\frac{\partial(\sin\theta A_\theta)}{\partial \theta} + \frac{1}{r\sin(\theta)}\frac{\partial A_\phi}{\partial \phi}$
Curl $\vec{\nabla} \times \vec{A}$	$\left(\frac{1}{r}\frac{\partial A_z}{\partial \phi} - \frac{\partial A_\phi}{\partial z}\right)\hat{r} + \left(\frac{\partial A_r}{\partial z} - \frac{\partial A_z}{\partial r}\right)\hat{\phi} + \left(\frac{1}{r}\frac{\partial(rA_\phi)}{\partial r} - \frac{1}{r}\frac{\partial A_r}{\partial \phi}\right)\hat{z}$	$\frac{1}{r^2\sin(\theta)}\begin{vmatrix} \hat{r} & r\hat{\theta} & r\sin\theta\,\hat{\phi} \\ \frac{\partial}{\partial r} & \frac{\partial}{\partial \theta} & \frac{\partial}{\partial \phi} \\ A_r & rA_\theta & r\sin\theta A_\phi \end{vmatrix}$
Laplacian $\nabla^2 f$	$\frac{1}{r}\frac{\partial}{\partial r}\left(r\frac{\partial f}{\partial r}\right) + \frac{1}{r^2}\frac{\partial^2 f}{\partial \phi^2} + \frac{\partial^2 f}{\partial z^2}$	$\frac{1}{r^2}\frac{\partial}{\partial r}\left(r^2\frac{\partial f}{\partial r}\right) + \frac{1}{r^2\sin\theta}\frac{\partial}{\partial \theta}\left(\sin\theta\frac{\partial f}{\partial \theta}\right) + \frac{1}{r^2\sin^2\theta}\frac{\partial^2 f}{\partial \phi^2}$

References

Griffiths, David J. (David Jeffery), 1942, Introduction to Electrodynamics. Boston: Pearson, (2013).

Jackson, John David, 1925–2016. Classical Electrodynamics. New York: Wiley, (1999).

Boas ML. Mathematical Methods in the Physical Sciences. 2nd ed. Wiley; (1983).

George B. Arfken, Hans J. Weber, Mathematical Methods for Physicists (Fourth Edition), Academic Press, (1995).

Mark A. Heald and Jerry B. Marion, Classical Electromagnetic Radiation, Third Edition (2013).

A. W. Joshi, Matrices and Tensors in Physics (1995).

Zhu, Y. Legendre Polynomial in equations and Analytical tools in Mathematical Physics, Springer (2021).

Anupam Garg, Classical Electromagnetism in a Nutshell, Princeton University Press, (2012).

Carl S. Helrich, The Classical Theory of Fields: Electromagnetism, (2012).

Aharoni, J., The Special Theory of Relativity, 2nd edition, Oxford University Press, Oxford (1965).

Barut, A.O, Electrodynamics and Classical Theory of Fields and Particles, Macmillan, New York (1964); Dover reprint (1980).

Anderson, J.L., Principles of Relativity Physics. Academic Press, New York, (1967).

Arzelies, H, Relativistic Kinematics. Pergamon Press, Oxford, (1966).

Byerly, W. E. Fourier Series and Spherical Harmonics. Ginn, Boston (1893).

Churchill, R. V., and J. W. Brown. Fourier Series and Boundary Value Problems, 5th edition. McGraw-Hill, New York (1993).

Jeans, J. H. Mathematical Theory of Electricity and Magnetism, 5th edition. Cambridge University Press, Cambridge (1925); reprinted (1958).

Jefimenko, O. D. Electricity and Magnetism. Appleton-Century Crofts, New York (1966); 2nd edition, Electret Scientific, Star City, WV (1989).

Johnson, C. C. Field and Wave Electrodynamics. McGraw-Hill, New York (1965).

Jones, D. S. The Theory of Electromagnetism. Pergamon Press, Oxford (1964).

Kilmister, C. W. Special Theory of Relativity, Selected Readings in Physics. Pergamon Press, Oxford (1970).

Lighthill, M. J. Introduction to Fourier Analysis and Generalized Functions. Cambridge University Press, Cambridge (1958).

GPSR Compliance
The European Union's (EU) General Product Safety Regulation (GPSR) is a set
of rules that requires consumer products to be safe and our obligations to
ensure this.

If you have any concerns about our products, you can contact us on

ProductSafety@springernature.com

In case Publisher is established outside the EU, the EU authorized
representative is:

Springer Nature Customer Service Center GmbH
Europaplatz 3
69115 Heidelberg, Germany